Springer Theses

Recognizing Outstanding Ph.D. Research

For further volumes:
http://www.springer.com/series/8790

Aims and Scope

The series "Springer Theses" brings together a selection of the very best Ph.D. theses from around the world and across the physical sciences. Nominated and endorsed by two recognized specialists, each published volume has been selected for its scientific excellence and the high impact of its contents for the pertinent field of research. For greater accessibility to non-specialists, the published versions include an extended introduction, as well as a foreword by the student's supervisor explaining the special relevance of the work for the field. As a whole, the series will provide a valuable resource both for newcomers to the research fields described, and for other scientists seeking detailed background information on special questions. Finally, it provides an accredited documentation of the valuable contributions made by today's younger generation of scientists.

Theses are accepted into the series by invited nomination only and must fulfill all of the following criteria

- They must be written in good English.
- The topic should fall within the confines of Chemistry, Physics, Earth Sciences, Engineering and related interdisciplinary fields such as Materials, Nanoscience, Chemical Engineering, Complex Systems and Biophysics.
- The work reported in the thesis must represent a significant scientific advance.
- If the thesis includes previously published material, permission to reproduce this must be gained from the respective copyright holder.
- They must have been examined and passed during the 12 months prior to nomination.
- Each thesis should include a foreword by the supervisor outlining the significance of its content.
- The theses should have a clearly defined structure including an introduction accessible to scientists not expert in that particular field.

Thomas Harry Sharp

Biomolecular Imaging at High Spatial and Temporal Resolution *In Vitro* and *In Vivo*

Doctoral Thesis accepted by
the University of Bristol, UK

Springer

Author
Dr. Thomas Harry Sharp
School of Chemistry
University of Bristol
Bristol
UK

Supervisor
Prof. Derek N. Woolfson
School of Chemistry
University of Bristol
Bristol
UK

ISSN 2190-5053 ISSN 2190-5061 (electronic)
ISBN 978-3-319-35023-3 ISBN 978-3-319-02159-1 (eBook)
DOI 10.1007/978-3-319-02159-1
Springer Cham Heidelberg New York Dordrecht London

Softcover reprint of the hardcover 1st edition 2014

Printed on acid-free paper

Springer is part of Springer Science+Business Media (www.springer.com)

Parts of this thesis have been published in the following journal articles:

T.H. Sharp, M. Bruning, J. Mantell, R.B. Sessions, A.R. Thomson, N.R. Zaccai, R.L. Brady, P. Verkade & D.N. Woolfson. "Cryo-transmission electron microscopy structure of a gigaDalton peptide fiber of de novo design." *Proc. Natl. Acad. Sci. USA* (2012), 109, pp. 13266–13271.

E. Brown, J. van Weering, T. Sharp, J. Mantell & P. Verkade. "Capturing Endocytic Segregation Events with HPF-CLEM." *Methods in Cell Biology* (2012), 111, pp. 175–201.

J.R.T. van Weering, E. Brown, T.H. Sharp, J. Mantell, P.J. Cullen & P. Verkade. "Intracellular membrane traffic at high resolution." *Methods in Cell Biology* (2010), 96, pp. 619–648.

Supervisor's Foreword

Dr. Thom Sharp worked between my group and that of Paul Verkade in Biochemistry, in the area of developing new methods for cryoelectron microscopy. He had to master a whole range of techniques and develop new skills before he could even start "researching": these included modelling, chemistry, molecular biology and microscopy; indeed, his examiners were astonished that he took all of this on himself. Thom used these skills to tackle an outstanding problem in my lab, and indeed in the protein materials field in general; namely the pursuit of a high-resolution structure of one of our peptide-based materials. Honestly, I believed that this could not be done. But Thom was keen, insistent and determined enough to have a go. He had to learn more methods—indeed, he had to develop new methods for analysing and processing this particular type of electron microscopy data, and in areas outside mine and Paul's expertise. Again, his examiners were impressed by this ability and tenacity. This effort of about a year resulted in the high-resolution structure (near-atomic resolution) of a peptide-based material that contains many millions of molecules. For the first time, this gives us a molecular description of a peptide-based material that we have developed over more than a decade. We published this work in *PNAS USA* in 2012. I suspect that it will have a huge impact in the peptide assembly field, and also in electron microscopy as it introduces new methods and approaches, all of which are of Thom's invention.

Bristol, August 2013 Prof. Derek N. Woolfson

Abstract

This thesis covers two separate projects linked by transmission electron microscopy (TEM) of biological samples: first, the utilisation of cryoTEM—TEM performed at cryogenic temperatures—to elucidate the superstructure of self-assembled peptide fibres, and second, the development of new probes for Correlative Light Electron Microscopy (CLEM) both *in vitro* and *in vivo*.

CryoTEM involves imaging samples in a hydrated state as close to experienced in bulk solvent as possible. This yields images and data that reflect the native state, and different from those of negatively stained samples that have been dried down and stained. Self-assembling peptide fibres (SAFs) were developed in the Woolfson lab in 2000 and characterised as α-helical coiled coils that grow both longitudinally and laterally. To date, data suggest an ordered superstructure, but negative stain TEM images and fibre X-ray diffraction have yielded only low resolution (~20 Å) information. In this thesis, I show that SAFs subjected to rapid plunge freezing and cryoTEM display a remarkable superstructure. Individual micrographs give high-resolution data that allowed direct structural interpretation of the packing of individual α-helices within the fibre, and the construction of a 3D electron density map at 8 Å resolution. Furthermore, an all-atom model was derived combining the cryoTEM data and a 2.3 Å X-ray crystal structure of a variant of the building block incapable of forming fibres. Together, these provide the highest-resolution structure of a *de novo* designed protein-based supramolecular fibre.

Green fluorescent protein (GFP) has revolutionised molecular and cell biology by enabling visualisation of endogenous proteins within live cells, providing information about protein motility, localisation and interactions. However, the resolution of the light microscope (LM) is inherently limited by the wavelength of visible light, a problem that can be overcome by EM. However, EM requires a fixed cell. Correlative light electron microscopy (CLEM) combines the advantages of both techniques to allow visualisation of proteins in the live cell prior to detailed analysis at high resolution in the EM. At present there is no genetically encoded monomeric protein that is visible in both LM and EM. Therefore, the aim of the second project was to design and characterise a protein that was both fluorescent and electron dense. To achieve this, concatenation of a metal-binding protein, metallothionein (MT), was used to cluster heavy-metal ions into electron-dense nanoparticles. This was fused to Enhanced-GFP (EGFP) and characterised as a general-purpose "clonable" tag for use in CLEM. Metal binding capabilities were probed with mass spectrometry and dynamic light scattering. *In vitro* localisation of the probe in the EM was achieved by encapsulation within liposomes and fusion to SNX1, a membrane tabulating protein. *In vivo*, the probe was fused to various intra- and extracellular localisation domains and whole cell CLEM attempted. The probe proved difficult to visualise, even after autometallography with gold. So, a separate complementary tag was also designed. A mutated intracellular protein, FK506 Binding Protein (FKBP12(F36V)), was fused to EGFP. FKBP12(F36V) has high affinity for a novel ligand (SLF′), which can be linked to a single gold particle and delivered to the cytoplasm. This would yield a fluorescent protein with an electron-dense particle noncovalently bound. Preliminary studies are described towards this second goal, and results so far appear promising.

Bristol, January 2012 Thomas Harry Sharp

Acknowledgments

Thank you Mum and Dad, for support, encouragement, endless food (Mum) and endless discussions (Dad), as well as Nathan for simply being a great older brother. I couldn't have hoped for a better family to belong to.

Thank you to Downham Market VI form and everyone there, teachers and friends alike; moving there was the best decision I ever made. Thank you Bristol for enabling me to find myself amongst such clever and funny people as a naïve 18-year old could wish for.

Thank you to my supervisors Dek and Paul, obviously, expectedly, but worth so much thanks for giving me the opportunity, freedom and gentle nudge in the right direction when needed.

Thank you to all the members of the Woolfson lab, past and present. I feel I should thank a few specifically; Aimee, for always being there when I needed someone to talk to and laugh with, as well as putting up with me, especially while I've been writing this thesis. Oh, and for the carbonara. Jordan, for the constant banter and the invention of the *real* cool-wall, as well as the rest of the lab for making the most stressful time of my life also one of the most fun: Nice one Craig, Tom V, Marc, Franziska, Drew for the Hammerhead Pigasus, Gail, Tim A, Kevin, Beth, Pat, Bertie, Daniel, Ed, and anyone else I've forgotten.

Out of the Woolfson lab and into the Verkade lab; Judith, Debbie, Gini and Big Ed have been a great source of knowledge and non-stop help. Jan, you're going to make a great boss one day, and an even better drummer.

Finally, thanks go as well to John, Rob, Malcolm, Marcus and Robbie for making the time after work so memorable.

Contents

Part II Materials and Methods

Abbreviations

AEM	Analytical electron microscope/microscopy
AFM	Atomic force microscopy
AFP	Autofluorescent protein
AuNP	Gold nanoparticle
AuTM	Aurothiomalate
BSA	Bovine serum albumin
BHP	Brandeis helical package
CCD	Charged-coupled device
CEMOVIS	CryoEM of vitreous sections
CLEM	Correlative light electron microscopy
CLSM	Confocal laser scanning microscope/microscopy
CryoTEM	Transmission electron cryo-microscope/microscopy
CTF	Contrast transfer function
DAB	Diaminobenzidine
DHB	2, 5-dihydroxybenzoic acid
DLS	Dynamic light scattering
EGFP	Enhanced green fluorescent protein
EDX	Energy-dispersive X-ray spectroscopy
EELS	Electron energy-loss spectroscopy
EM	Electron microscope/microscopy
EMDB	Electron microscopy data bank
EMS1	EGFP-MT2-SNX1 fusion protein
ERES	Endoplasmic reticulum exit site
FEG	Field emission gun
FKBP12(F36 V)	FK506-binding protein 12 kDa (Phenylalanine 36–valine)
FlAsH	Fluorescein bi-arsenical hairpin-binding
FLM	Fluorescence light microscope/microscopy
FRET	Förster (or fluorescence) resonance energy transfer
FT	Fourier transform
GA	Glutaraldehyde
GFP	Green fluorescent protein
GRAB	GFP-recognition after photobleaching
HPF	High-pressure freezing
IgG	Immunoglobulin-G

ILEM	Integrated light and electron microscope/microscopy
LM	Light microscope/microscopy
MALDI	MS Matrix-assisted laser desorption/ionisation mass spectrometry
MCS	Multiple cloning site
miniSOG	Mini singlet oxygen generator
MOPS	3-(N-morpholino)propanesulfonic acid
MT	Metallothionein
NTA	Nitrilotriacetic acid
PAG	Protein-A-gold
PALM	Photo-activated localisation microscopy
PDB	Protein databank
PFA	Paraformaldehyde
POI	Protein of interest
PSF	Point spread function
QD	Quantum dot
RBS	Ribosome binding site
ReAsH	Resorufin bi-arsenical hairpin-binding
SAF	Self-assembling fibre
SDCM	Spinning disk confocal microscope/microscopy
SEM	Scanning electron microscope/microscopy
SNR	Signal-to-noise ratio
STED	Stimulated emission depletion
STEM	Scanning transmission electron microscope/microscopy
STORM	Stochastic optical reconstruction microscopy
TEM	Transmission electron microscope/microscopy
TfnR	Transferrin receptor
TIRFM	Total internal reflection microscope/microscopy
WAX	Wide angle X-ray diffraction

Part I
Introduction

Chapter 1
Biomolecules and Microscopy

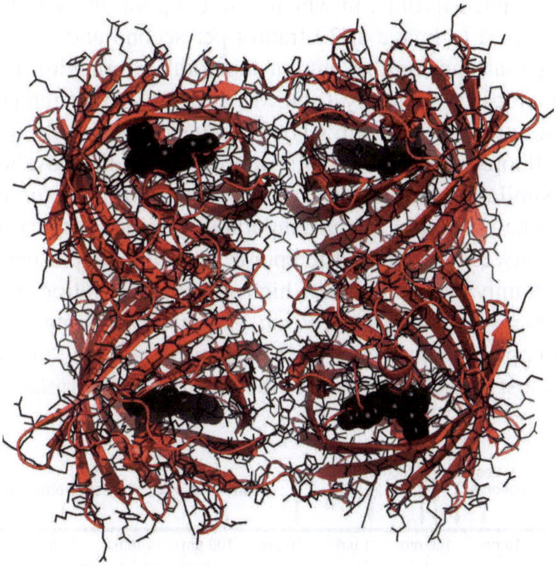

Biomolecular imaging has become key to visualising and understanding biological processes and systems. In this chapter focus is placed on the imaging aspect, specifically the difference between spatial and temporal resolution in relation to biomolecular systems. Light microscopy (LM) is particularly well-suited for imaging dynamic processes, such as those that occur within cells. The spread of both super-resolution LM and the use of genetically encoded fluorescent proteins is described in Sect. 1.2. Conversely, electron microscopy (EM) is typically used for high spatial resolution studies (Sect. 1.3). The use of EM to elucidate the structure of various protein fibres is described in Sect. 1.4. This chapter concludes with the combination of these two imaging modalities with a description of correlative light electron microscopy (Sect. 1.5) and the objectives of this Thesis (Sect. 1.6).

T. H. Sharp, *Biomolecular Imaging at High Spatial and Temporal Resolution In Vitro and In Vivo*, Springer Theses, DOI: 10.1007/978-3-319-02159-1_1,

1.1 Biomolecular Imaging

Biomolecular imaging is fundamental to understanding biological systems at high resolution, both spatial and temporal. Although an expansive definition, biomolecular systems are those which comprise, use, encode or modify biological molecules. Some examples of biomolecules include proteins and DNA, as well as steroids, haem groups and vitamins. Biomolecular systems span dimensions from the sub-nanometre to the metre regimes and include nanomachines, such as the ribosome, whole cells, tissues, and organisms. Imaging whole systems is therefore difficult to achieve using any single technique, although various methods are available for imaging ranges of scales (Fig. 1.1). For instance, magnetic resonance imaging (MRI) can be used to probe whole organisms, including animals and plants, with a spatial resolution of ~2–0.1 mm. However, spatial resolution is not necessarily all-important if the information is contained in time, in which case temporal resolution is paramount. MRI can now be used to image at 20 frames per second and, therefore, provides low-resolution spatial and high-resolution temporal information [1]. In contrast, spinning disk confocal microscopy (SDCM) offers a similar temporal resolution with a spatial resolution of ~200 nm; 10,000× better. However, the requirement that SDCMs require fluorescent tags and thin samples limits their use as *in vivo* diagnostic systems. Similarly, although the transmission electron microscope (TEM) can achieve sub-nanometre resolution, the temporal resolution is typically zero.

The tradeoff between spatial and temporal resolution at present resembles the Heisenberg uncertainty principle, where higher spatial resolution comes at the cost of lower temporal resolution and *vice versa*. As with everything, using the correct tool for the job is paramount. Or, as is increasingly the case, correlating the data output from the most suitable methods for the job. Indeed, "correlative microscopy"

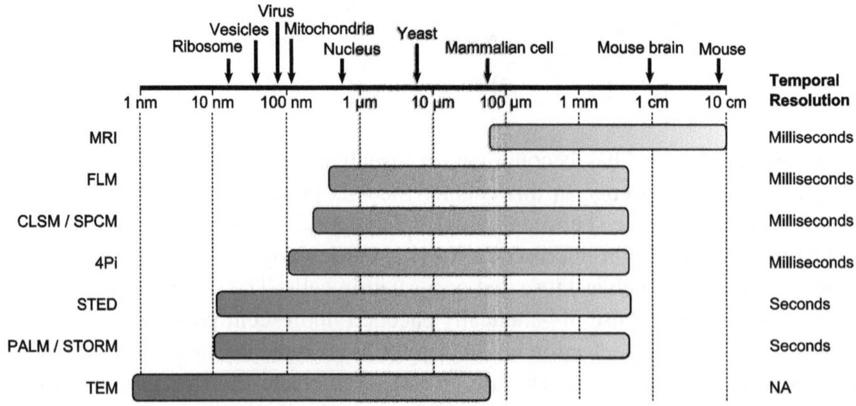

Fig. 1.1 Spatial versus temporal resolution. The spatial resolution is for lateral dimensions. Abbreviations: *FLM* fluorescence light microscopy; *CLSM* confocal laser scanning microscopy; *STED* stimulated emission depletion; *PALM* photoactivated localisation microscopy; *STORM* stochastic optical reconstruction microscopy; *NA* not applicable. Modified from [2, 3]

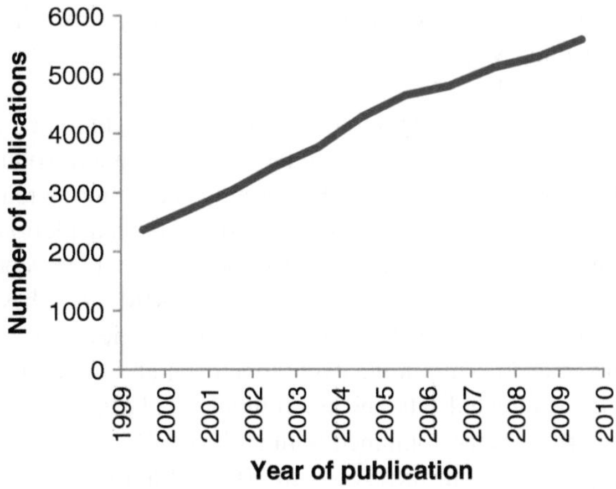

Fig. 1.2 The increasing interest in correlative microscopy. Results are from Google Scholar, searching both articles and patents for "correlative microscopy" between the dates given

is becoming increasingly pertinent in order to gather both high spatial and high temporal resolution data (Fig. 1.2).

1.2 Light Microscopy

Light microscopy (LM) allows live cells to be imaged non-destructively in real-time. However, the wavelength of visible light, from 380–750 nm, limits the resolution that can be achieved with traditional LM. Spatial resolution describes the minimum distance that can be resolved between two particles. For LM, one can discern between two points that are greater than half the wavelength of illuminating light, which is at best 200 nm apart, any closer and they appear as a single point. This is known as the Abbe diffraction limit. For comparison, a single mitochondrion or *Escherichia coli* bacterium is generally 2000 nm long and 500 nm in diameter, whereas the smallest organelles in eukaryotic cells range from 150 nm vesicles to 20 nm ribosomes, which are well below the limit of resolution.

Fluorescence light microscopy (FLM), imaging a chromatically labelled particle by re-emission of absorbed light, allows imaging of particles much smaller than 200 nm, but is still restricted by the limit of resolution. However, it has yielded a great deal of information on spatiotemporal relationships, such as protein co-localisation and motility. Indeed, using scanned light sheet microscopy has allowed subcellular detail to be imaged during the first 24 h of zebrafish embryogenesis [4]. Confocal laser scanning microscopy (CLSM) allows greater contrast of an image than widefield fluorescence imaging by using pinholes to eliminate out-of-focus light. However, the best theoretical resolution is still around 200 nm in the focal (x, y) plane, and only 800 nm in the axial (z) dimension [5]. Total internal reflection microscopy (TIRFM) has increased the axial resolving power of light microscopes by illuminating only a 100 nm evanescent field above the coverslip. This allows accurate analysis of dynamic events at or near the plasma membrane, such as vesicle fusion or microtubule dynamics [6].

Fluorescence light microscopy was thought to be limited by the Abbe diffraction limit. This is now known to be false, and superresolution systems—where this diffraction limit is surpassed—are now becoming increasingly common. Methods for superresolution, such as stimulated emission depletion microscopy (STED) [7], 4Pi microscopy [8], photo-activated localisation microscopy (PALM) [9] and stochastic optical reconstruction microscopy (STORM) [10], have achieved a resolution better than 50 nm. Indeed, STORM can now achieve a resolution of 20 nm in the focal plane and 60 nm in the axial dimension and therefore enable imaging of individual clathrin-coated pits and microtubules [11], much better than CLSM and TIRFM. Interferometry combined with PALM (iPALM) has since allowed imaging of fluorophores within fixed cells to 10×20 nm (in z and (x, y), respectively), nearing the resolution of electron tomography (see Sect. 1.3) [12]. STORM and PALM are comparable methods; fluorophores are activated one at a time and a point spread function is applied to determine the location with nanometre accuracy. The fluorophore is then turned off and another fluorophore imaged. Many rounds of this process reveals sub-diffraction images. The obvious requirement is photoswitchable or photoactivatable fluorophores. Some genetically encoded autofluorescent proteins (AFPs) based on green fluorescent protein (GFP) have recently been developed for such a purpose, such as the green-red photoconvertible protein Kaede [13] and the photoswitchable protein Dronpa [14]. Some AFPs, such as GFP [15] and reversibly switchable enhanced-GFP (rsEGFP) [16], can be directly imaged by STED, which rasters a laser in a manner identical to CLSM. The difference lies in that the excitation spot is overlaid with a doughnut-shaped function at a lower wavelength that de-excites the fluorophores, leaving molecules fluorescent only within a much smaller point spread function than standard CLSM. This mode of imaging is fast enough to be used to image dynamic events [17, 18] with a resolution in live cells of <80 nm [16]. Stimulated emission can even compete with non-radiative decay mechanisms and as such can be used to image non-fluorescent chromophores such as hemoglobin by collecting stimulation-emitted light from the de-excited molecules [19].

1.2.1 Green Fluorescent Protein

Cell biology has been revolutionised by recent advances in light microscopy and the development of fluorescent probes. Protein tagging with genetically encoded reporter constructs is the most reliable method of visualising endogenous proteins, and leading this revolution is green fluorescent protein (GFP).

GFP was isolated from the north Atlantic jellyfish *Aequoria victoria* by Shimomura et al. a discovery that earned him the joint Nobel prize for Chemistry in 2008 [20]. *A. victoria* contains Aequorin, a calcium-binding protein that emits blue light from a prosthetic group in a calcium-dependent manner. The emission peak

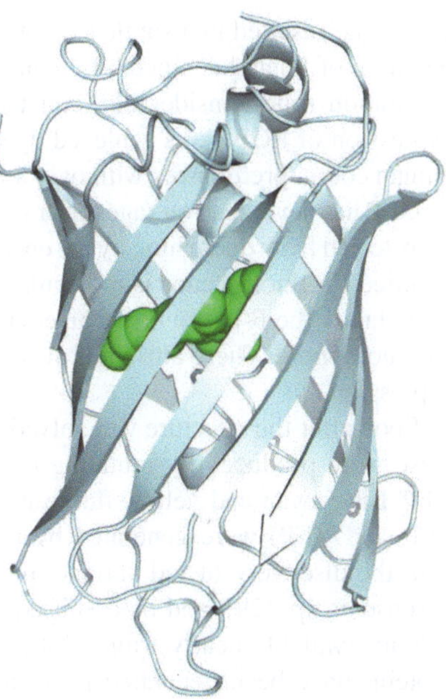

Fig. 1.3 The structure of GFP (PDB code 1EMA). The encapsulating β-can structure is drawn in cartoon form to allow visualisation of the fluorophore (*green*)

of Aequorin (470 nm) overlaps with the minor excitation peak of GFP (480 nm), which accepts the energy through radiation-less transfer and emits green light [21]. Aequorin is therefore the Förster (or fluorescence) resonance energy transfer (FRET) donor and GFP the acceptor.

The structure of GFP was solved by two separate groups in 1996 [22, 23]. GFP has a very stable β-can structure composed of 11 β-strands that encapsulate a single α-helix (Fig. 1.3). The fluorophore is located near the center of the β-can and is generated by auto-catalytic cyclisation of serine65, tyrosine66 and glycine67. Oxidation by molecular oxygen produces a planar π-conjugated system and converts GFP to the fluorescent product (Fig. 1.4) [24].

Initially, limitations of GFP included slow maturation, pH sensitivity causing a double excitation peak, and a tendency to aggregate. These drawbacks have been largely overcome by site-directed mutagenesis and codon-optimisation of the gene. Enhanced-GFP (EGFP) is a double mutant of wild-type GFP; mutation of Serine65–

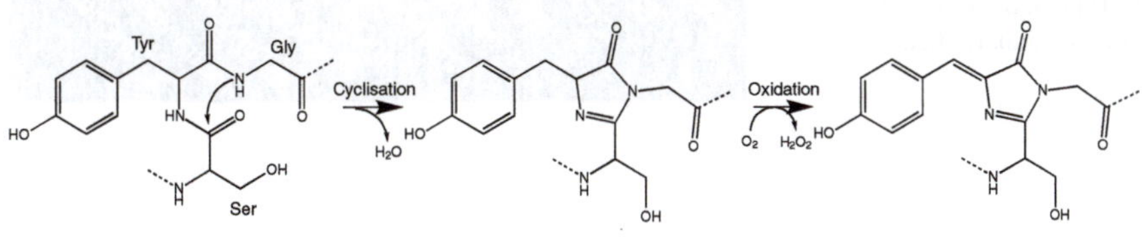

Fig. 1.4 Mechanism of formation of the GFP fluorophore. Modified from [24]

Threonine resulted in a single excitation peak and increased brightness [25], whilst mutation of Phenylalanine64–Leucine further increased the brightness and reduced maturation time considerably, concurrently reducing aggregation [26]. Increased expression of EGFP was achieved by codon-optimising the open reading frame for human codon preferences with over 190 silent base mutations [27].

GFP has several advantages over non-genetic fluorescent probes; the fluorophore is protected from quenching by the encapsulating protein, the only ancillary molecule required for fluorescence is molecular oxygen, photo-toxicity is minimal, and most importantly the β-barrel forms an environment ideal for auto-catalytic cyclisation of the fluorophore. These attributes make GFP an ideal fusion protein for heterologous expression.

Soon after the structure was solved, different coloured variants appeared. At first these were produced by mutating residues within or near to the chromophore of GFP. Blue, cyan and yellow fluorescent protein mutants (imaginatively titled BFP, CFP, and YFP) were generated by random mutagenesis of GFP [22, 28]. Then, with the discovery of red variants in non-fluorescent corals, such as DsRed from Discosoma sp. [29], and their subsequent subjection to mutagenesis, the palette of colours available nearly tripled [30] (Fig. 1.5). In subsequent years various novel proteins have been generated from these two basic subfamilies, including photo-activatable GFP (PA-GFP) [31] and photo-convertible GFP [14], some of which have found use in techniques for superresolution FLM.

The major disadvantage of methods for superresolution, as with all FLM techniques, is that only the fluorescent particles are visible and not the surrounding environment, which remains invisible. This so-called reference space is only visible with transmission microscopy. For LM this is usually phase contrast or differential interference contrast. However, for high-resolution ultrastructural data FLM must be correlated with transmission electron microscopy (TEM). Indeed, combining PALM with TEM has yielded information regarding localisation of single proteins within their ultrastructural environment, with much greater protein labelling density than immunolabelling combined with EM can achieve [9] (Fig. 1.6).

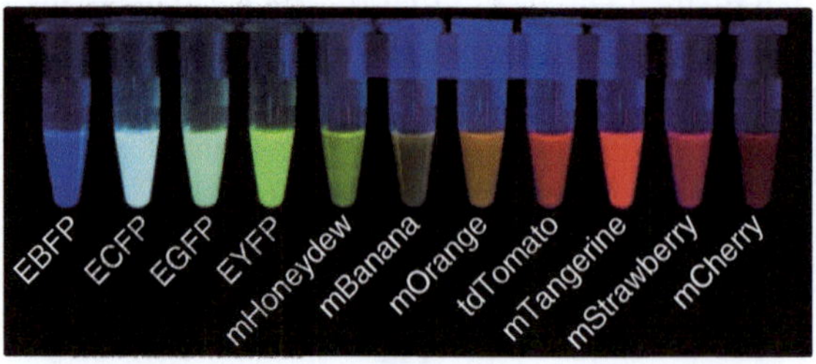

Fig. 1.5 Autofluorescent proteins evolved from GFP and DsRed. E stands for enhanced, m and td stand for monomeric and tandem dimer, respectively. Modified from the Tsien lab website (http://www.tsienlab.ucsd.edu)

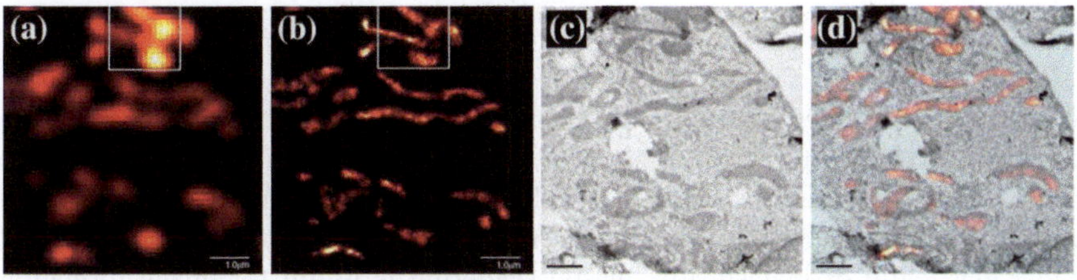

Fig. 1.6 A comparison between TIRFM, PALM and TEM imaging. **a** TIRFM, **b** PALM and **c** TEM images of the same location. **d** PALM overlays the TEM image yielding information on bulk protein localisation. Modified from [9]. Reprinted with permission from AAAS

1.3 Electron Microscopy

Since 1927 electrons have been known to possess wave-like properties, and in 1932 it was proven that electrons can be focused with magnetic lenses. The electron microscope (EM) rapidly followed these discoveries. The driving force for the invention of the EM was the limited resolving power of the LM, which until recently was limited to around half the wavelength of the illuminating light (see Sect. 1.2). Electrons have a wavelength much shorter than the distance between atoms, which are typically separated by ~1 Å (0.1 nm) in covalent bonds, thus implying that electrons can be used to image atomic detail. However, this limit has not been routinely achieved for biological specimens due to sample preparation and radiation damage, although the resolving power of EM can reach sub-nanometre distances, 1000× better than the LM.

This Thesis is primarily concerned with transmission EM (TEM); however there exist various other microscopy techniques including scanning EM (SEM), analytical EM (AEM) and scanning transmission EM (STEM), which are referred to within this Thesis. TEMs have a layout very similar to light microscopes, in that there is an illumination source, a condenser lens, objective lens, and an eyepiece. The most obvious difference is that the lenses used are electromagnets and not glass. Although all lenses suffer from abberations, those present in glass lenses can be minimized with the use of compound lenses. However, magnetic lenses suffer from various lens defects that cannot be corrected simply, described in detail below.

Perhaps the most important difference between TEM and LM is that the microscope operates within a vacuum. This is essential since electrons interact strongly with matter and scatter upon interaction with Coulombic forces, i.e., both the electron clouds and nuclei of atoms, and so an electron beam will not propagate in air. In order for a sample to survive insertion into the EM it must withstand the vacuum. Many inorganic crystals and particles do so intrinsically. Biological samples, however, must first be fixed so as to avoid explosive decompression or desiccation. Samples for TEM must also be thin enough for electrons to be transmitted with only one (or as few as possible) scattering events occurring. For specimens thinner than ~100 nm or single molecules this is not a problem, but thick cells and tissues must be sectioned prior to imaging.

1.3.1 Contrast

With TEM, the incident beam strikes the specimen and alters the transmitted beam in ways that produce contrast. Contrast is defined as a measurable change in amplitude of the electron waves. This is viewed as a change in intensity, which can be visualised with the aid of a phosphorescent screen or a CCD camera. Electrons are scattered either elastically or inelastically; elastic scattering is useful for image formation and electron diffraction, whereas inelastic scattering, although useful for AEM techniques such as energy-dispersive X-ray spectroscopy (EDX) or electron energy-loss spectroscopy (EELS), causes noise in TEM images and is therefore unwanted. The most common method of image formation is to block highly scattered electrons with an objective aperture, which can be exchanged or removed depending on the nature of the experiment and amount of contrast required, so areas which scatter electrons heavily appear dark (Fig. 1.7). This is the primary method of generating contrast for negatively stained specimens, although contrast may instead be generated using phase contrast [32].

1.3.1.1 Phase Contrast

Although a change in intensity is currently the only method to measure electron interactions, electrons obey wave-particle duality and therefore have a phase (θ) as well as an amplitude (A):

$$F(x) = A \cdot \sin(\omega x + \theta) \tag{1.1}$$

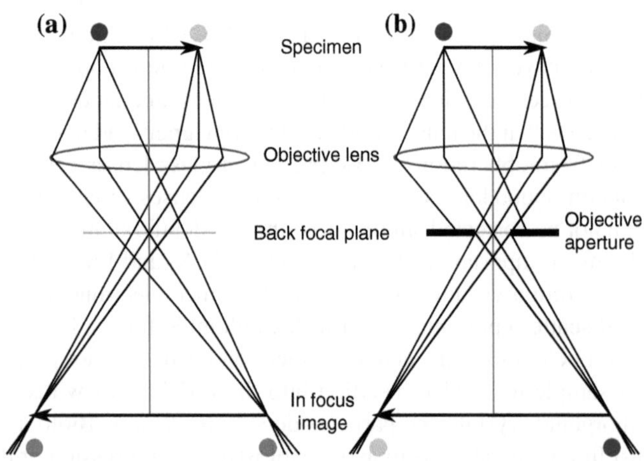

Fig. 1.7 Function of the TEM objective aperture. **a** Without an aperture, points that scatter electrons to high angles (*upper left spot*) and low angles (*upper right spot*) appear the same (*lower spots*). **b** With an aperture, highly-scattered electrons are intercepted, so the resulting point appears darker

Where ω is the frequency and assumed to be identical for coherent electrons. Phase information cannot be collected however, so any change in phase alone is impossible to image. Unfortunately, thin, unstained biological samples behave (ideally) as weak phase objects. That is, transmitted electrons carry information about the specimen in their phase only and do not change in amplitude. This means that if the image is in focus, the amplitudes of scattered and unscattered electrons are identical so there is no contrast and important information is lost. To overcome this, phase contrast microscopy is utilised in which the phases of the scattered electrons are forced to interfere with the transmitted electrons, either constructively or destructively, causing a change in amplitude. This is the mode of imaging with cryoTEM since there is no stain present. Generating phase contrast relies on the assumptions that the object is thin enough so that the electron waves do not change in amplitude within the specimen and that the incident beam is coherent. It is also dependent on the fact that magnetic lenses are imperfect and introduce phase shifts to the diffracted electrons.

1.3.2 Lens Defects and Resolution

There are various lens defects that act to limit the resolving power of the microscope. The three most common are spherical abberation, chromatic aberration and astigmatism (Fig. 1.8). Astigmatism can be corrected with stigmators, variable electromagnets that compensate for the defect, present in the condenser and objective lens systems. Chromatic aberration is reduced by using an electron source with as small as possible energy spread and having thin samples. However, since inelastic scattering increases the energy spread of transmitted electrons, chromatic abberation still causes a decrease in the information limit of the microscope. The information limit is equivalent to the resolution referred to for LM, where no spatial information is transmitted beyond this frequency (Fig. 1.9). To limit the effects of chromatic abberation, TEMs are available with correctors that selectively filter electrons based on their energy (wavelength) [33]. This is performed by placing an energy filter in the column of the microscope. The so called Ω-filter, due to its shape, selectively filters electrons based on their wavelength and passes specific energies through an aperture to allow image formation. Spherical aberration is the most important aspect of magnetic lenses since it limits the point resolution much more than chromatic aberration or astigmatism. The point resolution is the resolution to which the image can be directly interpreted without further processing and generally occurs at a much lower frequency than the information limit described above (Fig. 1.9). The effect that spherical aberration has on the image is complex, since there are effectively images of varying defocus superimposed at the "plane of least confusion" (Fig. 1.8a). This causes a point (P) at the imaging plane not to be brought to a point P' at the imaging plane, but is spread out to a disk at the plane of least confusion. This point has been affected by a point spread function (PSF) that influences the transfer of information through the objective lens. The process of image formation ($g(\mathbf{r})$) is therefore a convolution ($\otimes$) of the specimen ($f(\mathbf{r})$) with the PSF ($h(\mathbf{r})$), where $\mathbf{r}$ is the real space

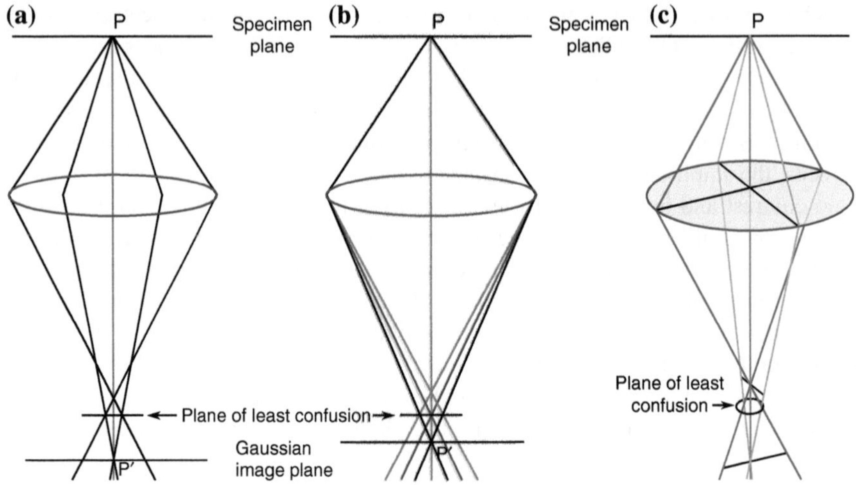

Fig. 1.8 Common lens defects in the EM. **a** Spherical abberation causes electrons emanating from a point (P) to be bent more the further from the center of the lens they pass. Paraxial rays focus to a point on the gaussian image plane (P′), whereas electrons that have been highly scattered by the specimen are focused to a point closer to the lens. **b** Chromatic abberation describes the phenomenon whereby electrons with different energies (wavelengths) are focused to differing degrees. Electron that have lost no energy (*black*) are focussed less strongly than those that have lost energy (*grey*). **c** Astigmatism is caused by a non-uniform magnetic field focussing electron differently around the lens and causes the image to appear stretched in different directions depending on the focus

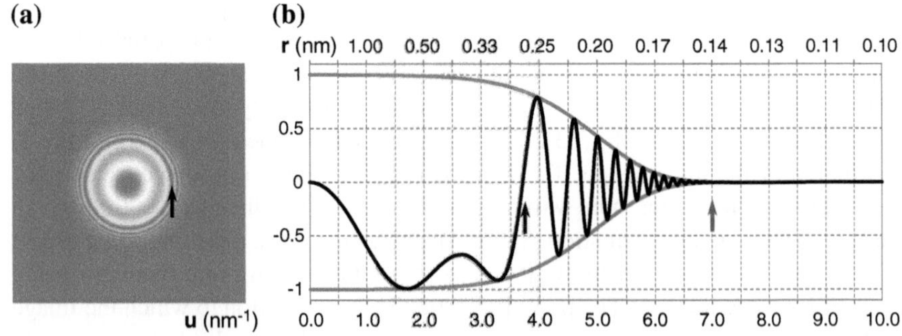

Fig. 1.9 Point resolution, information limit, and Thon rings shown for a theoretical EM at extended Scherzer defocus. **a** In reciprocal space, alternate Thon rings are phase flipped by 180°, with *dark circles* corresponding to the zeroes of the CTF. *Arrow* indicates the first zero of the CTF at 0.27 nm. **b** CTF (*black line*) and envelope function (*grey line*) showing the point resolution (*black arrow*) at 0.27 nm, and the information limit (*red arrow*). Data generated by CTF Explorer (http://maxsidorov.com/ctfexplorer/)

vector (notation has been adapted from [34]):

$$g(\mathbf{r}) = f(\mathbf{r}) \otimes h(\mathbf{r}) \tag{1.2}$$

$$G(\mathbf{u}) = F(\mathbf{u})H(\mathbf{u}) \tag{1.3}$$

In reciprocal space, which is viewed in the Fourier transform of an image, the PSF is more readily apparent as a series of circles known as Thon rings (Fig. 1.9). The position of the Thon rings is defined by the phase contrast transfer function (CTF) and is the main factor limiting information transfer and direct interpretation of the image. The phase CTF is characterised by oscillating phases; dark circles separating the Thon rings are where the phase flips from 0° to 180° and increase in frequency as the resolution increases [32]. The phase CTF ($\sin\chi(\mathbf{u})$) at any given spatial frequency (u) is affected by the defocus (Δf) and spherical aberration constant (C_s) of the objective lens, as well as the wavelength of the electrons (λ):

$$\sin\chi(\mathbf{u}) = \sin\left(\pi\Delta f\lambda u^2 + \frac{1}{2}\pi C_s\lambda^3 u^4\right) \tag{1.4}$$

The effects of spherical abberation can be partially compensated for by employing the Scherzer defocus (Fig. 1.9), which produces a wide bandpass of the same phase contrast. In other words, the Scherzer defocus increases the point resolution to the maximum value, yielding the most directly interpretable images possible with the microscope. This is also one of the definitions for the resolution of EM. Unfortunately, the Scherzer defocus also suppresses low-resolution information, so any large objects appear grey and are essentially invisible. The CTF therefore enforces a trade-off between contrast and resolution, which can be selected for by defocusing the objective lens.

The phase CTF affects image formation together with envelope functions that attenuate high resolution information (Fig. 1.9). Envelope functions derive from any aspect of the microscope that limits information transfer; these include the objective aperture, chromatic abberation, and specimen drift, among others. Together these form the complete CTF ($H(\mathbf{u})$). As mentioned, image formation by phase contrast is equivalent to convoluting the specimen with the PSF (Eq. 1.2), which is identical to multiplying the FT of the specimen ($F(\mathbf{u})$) with the CTF ($H(\mathbf{u})$, Eq. 1.3). When $\sin\chi(\mathbf{u})$ is negative, phase contrast is positive so atoms appear black on a white background, and *vice versa*. To increase the interpretable resolution of the images the phase CTF must be corrected, adding 180° to negative phase contrast rings to give white atoms on a dark background, which is a computational process and described below. However, recently C_s-corrected microscopes have become available [35, 36]. These have been developed by the materials-science community as an adjunct to image processing and are becoming more popular among the biological sciences [37]. Indeed, TEMs with both chromatic (C_c) and spherical (C_s) abberation correctors are now available with a sub-Ångstrom information limit [38].

A further development for TEM is the invention of phase plates able to generate phase contrast without employing a defocus [39]. Technologies developed for light microscopy, both the Zernike phase plate and a technique similar to differential interference contrast (DIC), have been adapted for EM. Zernike phase plates cause a phase shift of $-\frac{1}{2}\pi$ to the diffracted beam, which has already experienced a $-\frac{1}{2}\pi$ phase shift through diffraction. This places it $-\pi$ out of phase with the transmitted beam resulting in destructive wave interference and therefore positive phase contrast [40]. The disadvantage of this technique is the requirement of the electrons to be elastically scattered not only by the specimen, but also by the phase plate. However, several new designs for phase plates are being developed [41, 42], and should yield to biological problems in the near future.

With phase plates the point resolution is actually decreased with respect to the Scherzer defocus of standard TEM, although low spatial frequency information is transmitted with maximum contrast allowing both low and high resolution information to be visible at once. To circumvent the drawbacks of phase contrast imaging all of these modalities must be combined; the phase plate to increase contrast and the C_s-corrector to stop phase reversals of the CTF. An energy filter to remove inelastically scattered electrons would increase resolution further. For biological specimens, the EM also needs low dose imaging capabilities and a cryogenically-cooled specimen holder [37].

1.3.3 Fixation

Sample preparation for TEM involves protecting the specimen from the vacuum of the microscope. This can be done for biological samples a number of ways, primarily embedding in plastic and sectioning for large samples, whereas small biological macromolecules are usually negatively stained or frozen in vitreous ice.

Chemical fixatives, such as glutaraldehyde and osmium tetroxide (OsO_4), which cross-link proteins and lipids, respectively, introduce artifacts into the sample [43–45]. Chemical fixation takes from seconds to minutes to react and must diffuse into the cell, altering the intracellular environment as it does so. Cryo-fixation by high-pressure freezing (HPF) is regarded as the most reliable fixation method for whole cells [46]; it is much faster than chemical methods, taking on the order of milliseconds, and does not perturb the intracellular environment. The principle behind HPF is that at high pressure the nucleation of ice crystals is reduced, leading to the formation of vitreous ice. At 204.5 MPa ice forms at a density greater than that of liquid water (1.16 g cm^{-3} *cf.* 1.13 g cm^{-3}) and the freezing point of water is reduced to −22 °C. This allows greater penetration of freezing and substantially reduced damage to ultrastructure [47].

HPF avoids many of the pitfalls of chemical fixation, and can be performed on whole cells and even small blocks of tissue. However, thick samples cannot be viewed in the EM and frozen specimens therefore need processing. Freeze substitution is the method of choice for most cell biologists, where the frozen water is replaced with an organic solvent and fixed between −90 °C and −50 °C prior to infiltration with resin (reviewed in [48]). Here, the organic solvent (usually acetone) both dehydrates and fixes the sample, although extra fixatives are commonly added, such as OsO_4 and uranyl acetate to cross-link lipids and stabilise phospholipid head groups, respectively. This adds contrast to intracellular membranes that otherwise appear poorly contrasted [44].

If the sample is thin enough to be imaged directly then either negative staining with heavy metal salts or cryo-fixation may be achieved using a cryogen prior to imaging with cryoTEM. Alternatively, thick frozen hydrated samples can be sectioned prior to visualisation by cryoTEM, so called cryoEM of vitreous sections (CEMOVIS) [49], an extremely technically demanding step which adds artifacts peculiar to the technique. These include blade marks, chatters, and compression, although these can be somewhat reduced with an oscillating diamond knife [50]. Alternatively, focussed ion beam (FIB) milling can be used to create thin enough areas for cryoTEM [51]. However, even thick cells can be imaged directly if only the thin lamella around the edge is required for imaging. Keeping the sample constantly below the devitrification temperature of the specimen allows the ultrastructure to be viewed in a state as close to native as possible.

1.3.4 CryoTEM

1.3.4.1 Negative Stain and the Vitrification of Water

Negative stain involves embedding the molecule in a highly scattering matrix composed of heavy-metal salts (e.g., uranyl acetate, lead citrate, etc.) which infiltrates the sample to an extent that is dependent on the topology of the molecule. Heavy atoms scatter electrons to a greater extent than organic molecules—composed primarily of carbon, oxygen, nitrogen and hydrogen—and therefore create a "negative" image (Fig. 1.10). The obvious drawback is that the molecule is no longer hydrated or in its native state, indeed it has been adsorbed onto a flat surface, then covered with a heavy salt solution before being dried and placed in a vacuum. Nonetheless, this technique yields very important information regarding the molecules size, shape, topology and has been used to reconstruct 3D structures suitable for atomic resolution structure fitting [52].

Although negative stain allows easy visualisation of particles, the contrast is actually created by the stain, not the protein. In Fig. 1.10 on the left hand side is an image of a negatively stained T4 bacteriophage, which appears like a sac with a tail since the stain molecules have not infiltrated the head and have instead filled the collapsed regions, revealing topological information about the drying artifacts. On the right is a similar bacteriophage imaged by cryoTEM after vitrification by plunge freezing, with no stain present. The salient features are still recognisable; the size, shape and even tail striations. However, the head now looks like a uniform disk, caused by contrast being generated by the difference in densities between protein, DNA, and

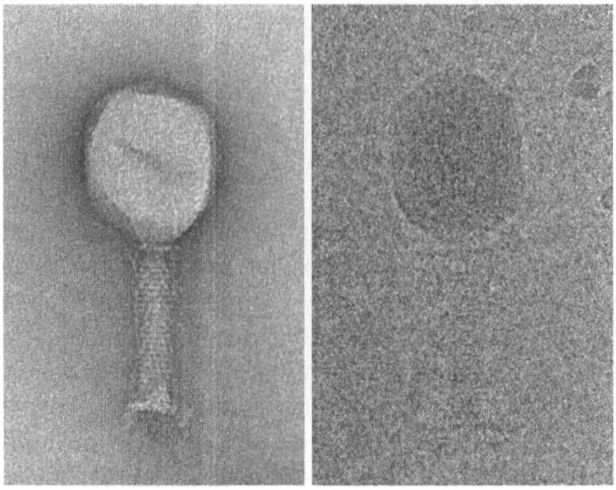

Fig. 1.10 T4 bacteriophage imaged after negative stain (*left*) or vitrification utilising phase contrast (*right*). Images taken by the author on a Tecnai T20

vitreous ice. The DNA coiled up within the head can readily be imaged by changing the focus, which is not achievable with negative stain. The other obvious differences are the high amount of noise and the low contrast of the cryoTEM image, caused by the requisite low-dose imaging method for frozen-hydrated samples.

CryoTEM was developed to avoid some of the pitfalls of negative stain. Biological molecules are nearly always found in aqueous suspensions, with the notable exception of lipid-bilayer membrane spanning proteins, and it is therefore desirable to image these molecules in amorphous water or buffer. The problem with such an approach is that water will explosively decompress in the vacuum of the microscope, destroying the sample. An obvious solution is to embed the molecule in ice and keep it frozen. However, ice formation is a process of crystallisation that perturbs the amorphous state of liquid water and results in phase separation of salts, creating osmotic and pH gradients and severely disrupting cells and protein molecules. Ice also takes up more space than liquid water, generating pressure forces throughout the sample. Crystalline ice also scatters electrons efficiently, meaning that in the microscope it is opaque or translucent. Pure water can be cooled down to −38 °C and remain liquid, so-called supercooling. However, as soon as this homogenous nucleation temperature is reached every water molecule can behave as a crystal nucleus and the water rapidly freezes to form hexagonal ice (the familiar species found on mountains and in freezers) [53]. As soon as crystallisation begins the latent heat of fusion released by crystallisation increases the temperature until crystallisation is complete. The only way to avoid hexagonal ice formation is to cool the water extremely quickly forming so-called vitreous ice, which due to this difficulty was discovered only 30 years ago [54]. This can only be achieved if the rate of heat withdrawal is faster than the release of the heat of fusion. The rate necessary to cool water to form vitreous ice is of the order of 10,000 K s^{-1} [55]. To reach these rates, the water must be immersed in a coolant that can extract heat efficiently (i.e., have a high thermal conductivity) from the water without increasing in temperature itself (i.e., have a high heat capacity). One such coolant is ethane which, with a melting temperature of −184 °C and a boiling temperature of −89 °C, remains liquid when in contact with above-room temperature water. The same is not true for liquid nitrogen which is subject to the Leidenfrost effect, where a film of insulating gaseous nitrogen forms around immersed surfaces and greatly reduces cooling rates. Liquid ethane has a specific heat capacity of 2.3 J g^{-1} K^{-1} (*cf.* $N_2(l)$: 2.0 J g^{-1} K^{-1}) and a thermal conductivity of 240 mJ m^{-1} s^{-1} K^{-1} (*cf.* $N_2(l)$: 153 mJ m^{-1} s^{-1} K^{-1}), making it an ideal liquid coolant (data extracted from [53]). Even if the rate of heat extraction is sufficient to produce vitreous ice, the internal flux of molecules within the sample rapidly becomes rate limiting since it takes heat longer to reach the coolant than be extracted by it. For this reason, samples can only be vitrified if they are thin (<500 nm).

Forming a stable film of water this thin and maintaining it long enough to be vitrified is usually achieved with the use of hydrophilic carbon films with many perforated holes (lacey-carbon and holey-carbon grids differ in the manner of preparation but not the problem they solve). The water extends across the hole and is held in a partially stable state by surface tension before the grid is plunged rapidly into liquid ethane. The use of holey-carbon film also allows the sample to be imaged suspended in vitreous ice, reducing the likelihood of imaging preferential orientations of the molecule since it is not adsorbed onto a surface. Continuous-carbon films also reduce the contrast of the image due to scattering from the carbon, which is eliminated with holey support films.

1.3.5 Image Acquisition

After freezing, the sample must not be allowed to heat up above the recrystallisation temperature, which for pure water is −143 °C but thankfully for cells is increased to ∼ −85 °C due to the cryoprotective effect of proteins and salts present. For imaging, the grid is placed in a cryoholder kept at ∼ −178 °C by $N_2(l)$.

The primary reason for using cryoTEM is that the structure of the molecule of interest itself is imaged, unlike negative stain which primarily shows surface features or topology (Fig. 1.10). It follows that any beam damage will perturb the molecule and degrade the information content of the resulting image. To reduce beam damage a low-dose method of imaging is used and the location of interest is subject to the minimum dose possible. Electron doses of typically ∼10 electrons $Å^{-2}$ are possible before damaging the sample, allowing the acquisition of only a few images [56, 57]. To minimise dosage, a three stage imaging procedure is used (Fig. 1.11). First, the area of interest is located at low magnification (and hence low dose). The focus mode is then engaged, where the beam is deflected to pass through an adjacent region where the focus is set, usually by imaging some of the carbon film or a sacrificial molecule. Finally the beam defection coils are switched off and the image acquired.

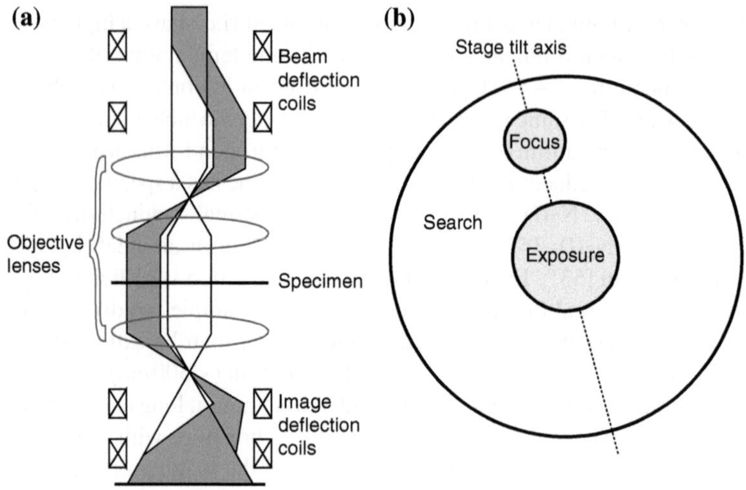

Fig. 1.11 Low dose imaging. **a** Microscope deflection coils enable focussing on an adjacent region without physically moving the stage. **b** The three areas used during low dose. The focussing area is generally on the tilt axis of the stage to minimise sample tilt artifacts, essential during tomography

1.3.6 Image Analysis

CryoTEM can image molecules in their native, fully-hydrated, state, unlike negative stained samples which are actually imaged as negative "shadows" of the molecule. However, since there is no stain present phase-contrast must be used to generate image contrast. This results in a compromise between resolution and contrast since contrast is generated by defocusing the objective lens. Generally, a compromise is reached whereby a series of images at different defocus values are taken before reconstructing the images computationally [58]. Similarly, if the molecules to be imaged are homogenous then different areas can be imaged sequentially. Images of molecules in the same orientation are then aligned, CTF-corrected, and averaged [59]. For very small particles simply finding them is difficult due to the low SNR. Therefore, low defocus (high resolution but low contrast) images can be taken prior to high defocus (low resolution but high contrast) images. The particles are then identified in the second image before their positions are used to extract the data from the high-resolution first image [60].

Reconstructing 3D models of proteins is the obvious successor to 2D views. Numerous methods have been developed to gather 3D information and reconstruct the target molecule. These include helical reconstruction [61], single-particle analysis [62], and tomography [63]. Such is the success of these approaches that hybrids of these individual techniques are now routinely used in dozens of labs around the world. These include single-particle analysis combined with helical reconstruction of filamentous structures, known as the iterative helical real space reconstruction method (IHRSR) [64]. Single-particle analysis has also recently joined traditional

2D electron crystallography for structure determination of membrane proteins by utilising lipid nanodisks [65]. These are small lipid bilayer disks stabilised by protein, into which membrane proteins can be inserted and treated as single particles for analysis. Electron tomography of chemically or cryogenically fixed cells can be combined with sub-tomogram averaging and pattern recognition to yield the contents of cells filled with individual protein molecules [66]. The software available to perform these reconstructions has also multiplied in recent years [67].

1.3.7 Helical Filaments

Helical filaments are unique in the respect that a single fibre provides different views of the constituent proteins, which obviates the need for tomography or single particle analysis [68]. Indeed, such is the usefulness of this approach that 2D crystals have been forced into helical arrays by crystallisation within the bilayer of lipid nanotubes to provide different views of the sample without the use of tomography [69]. The relative simplicity in gathering this information led to helical arrays becoming the first 3D structures solved by EM [68]. This was helped by the fact that information is constrained to layer lines in Fourier space, which makes the mathematical task of reconstruction much simpler.

A continuous helix can be described by two parameters; the radius (r) and pitch (P) (Fig. 1.12a). For discontinuous helices, two extra parameters are required; Δz and $\Delta\phi$, respectively the axial and angular translations of the subunits (Fig. 1.12b). Just like the two spatial vectors required to draw a 2D lattice, these provide the two vectors needed to draw a helical lattice (Fig. 1.13).

In reciprocal space, the FT of a helix is characterised by intensity along layer lines, the distance of which along the meridian can be used to determine Δz. However, $\Delta\phi$ cannot be determined directly. Each layer line corresponds to a new helix that can be drawn through the series of points that lie on the original helix (for two

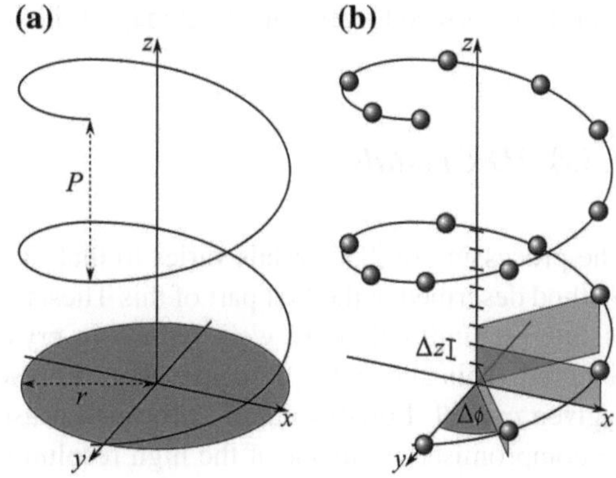

Fig. 1.12 Parameters for continuous and discontinuous helices. **a** A continuous helix can be described by the pitch (P) and radius (r). **b** A discontinuous helix can be described by adding Δz and $\Delta\phi$, which define the translation and rotation between subunits. Redrawn from [70]

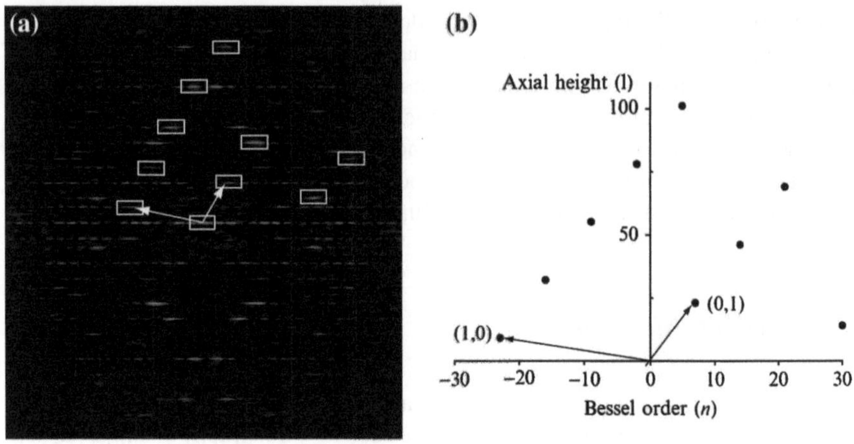

Fig. 1.13 Layer lines and Bessel functions. **a** FT of Ca-ATPase crystals showing layer lines (*horizontal streaks*) after indexing the near side (*white boxes*). **b** Miller indices assigned with an n,l plot of Bessel order versus layerline height, showing how a helical lattice can be represented as a 2D lattice. Modified from [70]

excellent reviews, see [71, 72]). The start-number of the helix of each layer line can be calculated with the use of cylindrical Bessel functions, which describe the distance from the meridian of the first major peak in the FT. The distance from the meridian increases with Bessel order, and since they are cylindrically averaged they are reflected in both the meridian and equator, generating the X pattern famous from the B-form DNA diffraction pattern. This reflection is due to both the near and far side of the helix being visible at the same time. The start-number of the helix is the same as the order of the Bessel function. So, from the Bessel order (n) and layer line height (l) both $\Delta\phi$ and Δz can be determined, respectively. In real space, each layer line and Bessel function appear as density waves around a cylinder. Summing these waves, much like for a 2D lattice (see below), gives the electron density around the cylinder. Therefore, from these parameters the helical lattice can be calculated and back-transformed to generate a 3D map of the helix [70].

1.3.8 2D Crystals

The processing of 2D crystals varies to that of helical filaments and is the primary method described in the first part of this Thesis. Since the locations of the molecules of interest (unit cells) are well defined in crystals, long range information is not required. This allows the defocus value to be set to the most convenient value for a given crystal. Low-resolution information used to find the crystal can therefore be compromised in favour of the high resolution used to elucidate the structure of the molecules, removing somewhat the need for a defocus series. This is because the defocus can be chosen so that the Thon rings coincide with the location of the diffraction maxima. Indeed, forcing single particles into 2D arrays has shown promise in increasing the resolution of averaged 2D images of proteins with a decreased number of particles [73].

In the processing of images of 2D crystals there are several assumptions that can be used to improve the resolution of the final image [74]. These will be described briefly in the order they are encountered in the processing.

Protein crystals can be thought of as a convolution of the protein and crystal lattice, as though a "protein-shaped brush" has been used to draw the crystal lattice points. The FT process converts spatial information into the frequency domain. Simply put, any image can be produced by adding together sine waves of different directions, amplitudes, phases and wavelengths (a Fourier summation or series, Fig. 1.14). A Fourier transform simply expresses the image as the direction, intensity and frequency components of the sine waves. The phase information is also present, but not usually displayed. Since crystals obey Bragg diffraction the direction of constructive interference of the waves is restricted to certain angles, which manifests in the FT as points of high intensity (i.e., a diffraction pattern). As stated above for the PSF, convolutions in real space are identical to multiplications in reciprocal space, so the image is formed by multiplying the FT of a single protein molecule with the FT of the crystal lattice. This means that the diffraction pattern is sampled at only the lattice points which are non-zero. In reality, however, noise permeates the image (due to inelastic electron scattering) and is therefore found across the entire diffraction pattern. One of the advantages of working with crystals is that, if the crystal lattice is known, all the signal present outside of the lattice points can be discarded by masking the holes of the diffraction pattern (so-called Fourier filtering). This greatly reduces the noise present in the filtered image. Certain 2D crystal space groups can also limit the variations in the diffraction pattern, allowing certain reflections to be averaged further reducing noise. However, this Thesis is concerned with protein crystals in the

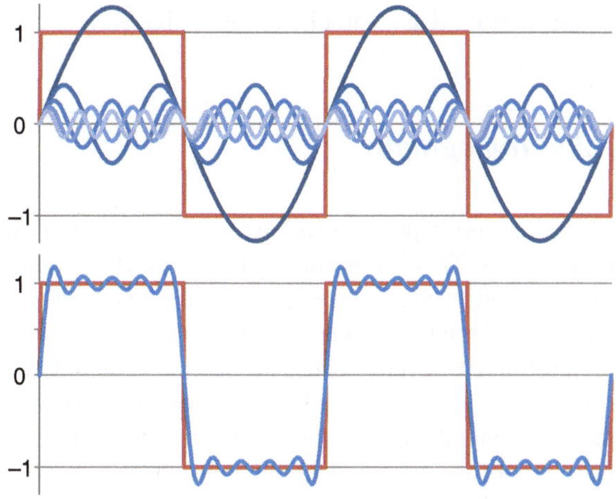

Fig. 1.14 Fourier series of sine waves. *Top*, constituent waves. *Bottom*, summed waves. Target waveform is in *red*, successive harmonics are in lighter shades of *blue*. The first five terms of $f(x) = \frac{4}{\pi}\sum_{\omega=1}^{\infty}\frac{\sin((2\omega-1)x)}{(2\omega-1)}$ are shown

p1 space group, which places no restriction on the diffraction pattern and therefore no further mention of crystal symmetry will be made.

Images of 2D crystals are subject to various distortions due to sample preparation, imaging such as beam induced movement or damage, and lens defects. This smears out the diffraction spots into broader peaks. After image acquisition such distortions can be reduced by "unbending"; computationally moving the unit cells to their expected locations so as to improve the crystal regularity and reduce FT artifacts [75].

The penultimate step is correcting the CTF. Certain reflections will have been phase flipped by 180° due to the spherical aberration and defocus of the objective lens [32]. The CTF can be defined by searching for the zero-intensity circles between the aforementioned Thon rings, where the information transfer is zero (the phase flips from 0° to 180°). It is then a simple matter of adding 180° to the phases of reflections in alternate rings to bring all the reflections into the same phase contrast. Whether this is to odd or even rings depends on whether the images are of negatively stained (negative phase contrast) or frozen-hydrated (positive phase contrast) samples. It is important to note here that the CTF is a further multiplication of reciprocal space, so where the CTF has zero-intensity the diffraction pattern will also have zero-intensity. This can result in missing reflections and has a major influence on the amount of defocus chosen for imaging. The intensity of the reflections is also modified by the CTF.

Finally, a map of the ideal unit cell is created by the Fourier summation of the sine waves described by each reflection now that they have been filtered and CTF-corrected. With non-p1 space groups, symmetry-related reflections are also averaged as described above before summation.

To reduce noise artifacts further, images can be merged by averaging their reflection amplitudes and phases [76]. This requires setting the center of the unit cells (the phase origin) by modifying the phases by certain amounts in x and y so that the phases are as close as possible to each other between images (reducing the phase residuals). This assumes that the crystals are identical and have the same orientation, an assumption that cannot be made in all cases.

1.3.9 Tomography

Electron tomography (ET) allows the generation of 3D volumes from a series of 2D images [77]. This is possible by backprojection; aligning each image to a common tilt axis and projecting the electron density back to this axis. This increases the resolution from the thickness of the section or sample to around 10 nm in the lateral dimension. However, because the sample is mounted within an EM grid it cannot be rotated around 360°. Indeed, because of the incident angle of electrons, a tilt greater that 70° generally causes too much ice to be in the lateral direction. Also, due to the geometry of the objective lens and obscuration by the EM grid itself, the greatest tilt range is ±70°. This leaves a "missing wedge" of information, leading to streaking

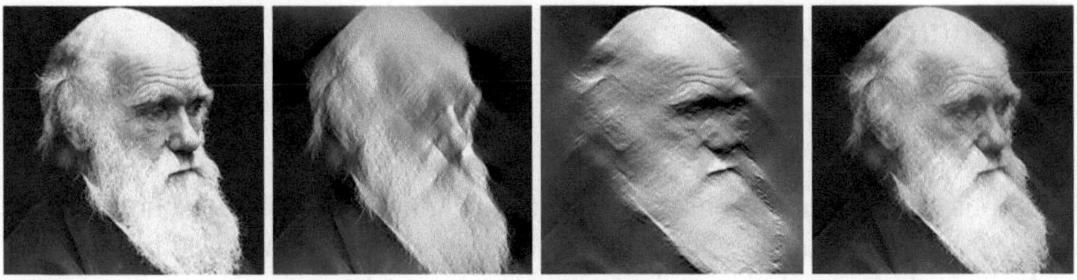

Fig. 1.15 Consequences of the missing wedge of tomography. From *left* to *right*: Original image; ±50° tilt from the *horizontal* axis; ±50° tilt from the *vertical* axis; averaging the two tomograms

of the electron density in the direction of the missing data (Fig. 1.15). Nonetheless, combined with frozen hydrated samples, cryoET provides nanometre resolution of native-state biomolecules.

CryoET presents its own problems, one of which is that electron doses must be much lower than for single images. Indeed, the total dose for a single image must be spread over the whole tomogram to prevent damage to the specimen, resulting in doses of ~1 electron $Å^{-2}$ for each image [57]. To circumvent these drawbacks, specific macromolecules within cells can be extracted and subject to sub-tomogram averaging to fill in the missing wedge of information and increase the resolution. Combined with pattern recognition, cryoET can yield the contents of cells filled with individual biomolecules [66, 78].

1.4 Protein and Peptide Fibres

Filamentous proteins are ubiquitous in Nature. They are found across all kingdoms of life, where they occur within cells, attached to cell surfaces and extracellularly. They perform a wide variety of cellular functions, including giving shape and stability to cells and tissues, acting as highways for intracellular transport and producing glutinous materials for cell attachment. In prokaryotes the flagellum enables movement of cells through aqueous media, whereas in eukaryotes actin filaments provide the machinery underpinning cell movement. All of the fibres within Nature are composed of smaller monomeric building blocks, typically of nanometre dimensions and kilo-Daltons in mass. These assemble through spontaneous or assisted processes to form fibres large enough to span distances of tens of microns. To gain an understanding of the assembly of these systems requires knowledge of their structure, both before and after polymerisation. Whilst globular subunits are more amenable to crystallisation and X-ray diffraction than the fibrous assemblies, currently the only way to solve the structures of filamentous structures is by TEM, usually combined with X-ray fibre diffraction and modelling. This section focusses on the advances made with the use of TEM.

1.4.1 Natural Fibres

1.4.1.1 Eukaryotic Fibres

Eukaryotic intracellular fibrous components, including actin polymers, intermediate filaments, and microtubules, are amongst the best-studied examples at the structural level, although the resolution and manner of structure determination varies.

Actin exists in an equilibrium of globular and fibrous states (G-actin and F-actin, respectively). F-actin is a homopolymer of G-actin which binds with the essential cofactor ATP. The structure of G-actin has been difficult to solve due to its propensity to form polymers. Structures have been solved for modified monomeric G-actin by inhibiting polymerisation, for example by covalent modification of a surface-accessible cysteine residue [79] or mutagenesis of key residues [80]. These studies showed a bilobed structure, each with 2 subdomains (numbered 1–4, Fig. 1.16) with the nucleotide binding domain in a cleft between the two lobes and accessible to each of the 4 subdomains (Fig. 1.16).

The formation of F-actin relies on ATP binding to G-actin, which then binds to the growing ends of F-actin, preferentially the barbed (+) end. Although G-actin is a poor ATPase, F-actin is an efficient one and filaments contain mostly ADP-actin [83]. Hydrolysis of ATP is fast, but release of the inorganic phosphate (P_i) is much slower. Whilst P_i is bound, F-actin is stable. As soon as it is released, however, disassembly of the filament occurs preferentially from the pointed (−) end. This transient nature of F-actin makes structural characterisation of the filament difficult, since crystals suitable for X-ray diffraction are unstable over the timescales of formation. However, cryoTEM has been used to obtain progressively higher resolution structures up to the 6.6 Å electron density map reported by Fujii et al. [82] and the 6 Å map from

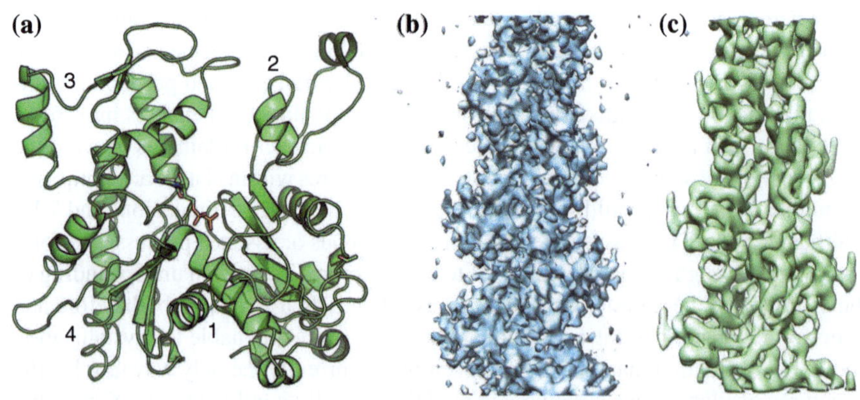

Fig. 1.16 The structures of G- and F-actin. **a** G-actin (PDB entry 1J6Z) is subdivided into 2 lobes, each with 2 subdomains (1–4 as numbered) with one molecule of ADP between them (shown as sticks) [79]. F-actin solved to 6 Å [81] **b** (EMDB entry 1674) and 6.6 Å [82] **c** (EMDB entry 5168). The barbed (+) end of actin is at the *bottom*

Murakami et al. [81] (Fig. 1.16). Key to their success was a TEM equipped with a FEG combined with in-column Ω energy filter (see Sect. 1.3.2). This increased the visibility of the otherwise low contrast 6 nm wide filaments. Both groups used methods developed for single-particle analysis to align and merge the data before imposing helical symmetry on the resulting high-resolution maps. However, due to noise, about 5% of the images were discarded due to difficulty in determining the orientation of the helix.

Crystal structures could be fitted into the electron density maps and predictions made about filament formation, ATP hydrolysis and nucleotide release. The two maps, although differing in their method of generation and in the crystal structures used for model building and fitting, did agree on the structural changes of G-actin within the filament; specifically, flattening of the bilobed structure and widening of the cleft between domains 1 and 3. These movements cause rearrangements in the chains surrounding the nucleotide binding cleft, leading to hydrolysis preferentially in F-actin. However, the two disagree on the biochemical implications of this structural change, with Fujii et al. suggesting Glu137 becomes nucleophilic [82], whilst Murakami et al. suggest a water molecule is rendered nucleophilic *via* a hydrogen bonded network to His161 [81]. These predictions can now be tested to discover which is correct.

Microtubules are similar to actin in that they also assemble in a nucleotide-dependent manner, this time with GTP as a cofactor. The tubules themselves consist of longitudinally repeating $\alpha\beta$-tubulin dimers, known as protofilaments, that associate laterally to form long hollow tubes. The α and β-tubulin molecules share both sequence and structural homology and as such are almost indistinguishable beyond ~8 Å resolution [84]. The structure of the $\alpha\beta$-tubulin dimer could not be solved by standard crystallographic techniques since the dimers associate longitudinally and curve to form spirals and circles, prohibiting crystallisation. However, tubulin protofilaments can be induced to form flat sheets by incubation with zinc and must be stabilised with Taxol [85]. This allowed the structure of the $\alpha\beta$-tubulin dimer to be solved by electron crystallography, first to 3.8 Å [85] and later refined to 3.5 Å [86] (Fig. 1.17).

Due to the difficulty in discriminating between the tubulins, cryoTEM has not identified high-resolution features of the filaments, although it has allowed certain structural constraints to be applied. Microtubules generally contain 13 protofilaments arranged in a ring which adopt a 3-start helix (Fig. 1.17). This results in a "seam", along which repeating lateral contacts between identical tubulins (i.e., α to α and β to β) cease and α to β contacts occur. This was thought to be a mechanism of depolymerisation, where microtubules peeled apart along the seam. CryoTEM, however, allowed Sui et al. to image microtubules with varying numbers of protofilaments with differing helix start-numbers, which has shown that this is not the case [87]. Sui et al. had access to a 400 kV cryoTEM which has a better resolution than 200 kV TEMs used for imaging actin by virtue of the higher accelerating voltage [87]. High energy electrons also suffer fewer inelastic collisions, decreasing the noise of the images. Using these images to reconstruct 3D electron density maps containing differing numbers of protofilaments allowed the docking of the 3.5 Å structure. This gave

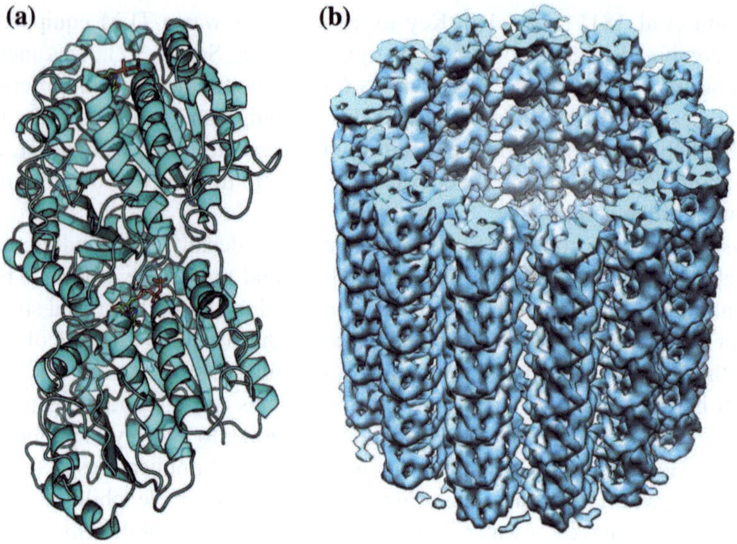

Fig. 1.17 Structures of the $\alpha\beta$ tubulin dimer and microtubule. **a** The $\alpha\beta$-tubulin dimer (PDB entry 1JFF) solved to 3.5 Å by electron crystallography [86]. The structural similarity between α-tubulin (*bottom*) and β-tubulin (*top*) is clearly visible here. GTP and GDP (shown as sticks) are bound to α and β-tubulin, respectively. **b** Microtubule containing 13 protofilaments solved to 9.8 Å [87] (EMDB entry 5193)

plausible predictions pertaining to inter-filament interactions, which were found to be primarily electrostatic and conserved between α- and β-tubulin. However, the authors provide no explanation as to how the fibres were oriented relative to one another prior to merging the data.

Of more direct relevance to this Thesis, intermediate filaments (IFs) are assemblies of α-helical coiled coils and, in contrast to actin and microtubules, are non-polar. IFs have proven intransigent to detailed structural studies due to their inhomogeneity; there are at least 65 separate genes for IFs, separated into 3 assembly groups within which IFs can form from heteropolymeric assemblies (reviewed in [88]). IFs comprise keratin, lamin and vimentin, as well as numerous other filamentous structures. They are thought to assemble through a 5-stage process involving; (1) dimerisation to form parallel α-helical coiled coils, (2) formation of an antiparallel dimer of dimeric coiled coils that renders them non-polar, (3) unit length filament (ULF) formation of 8 tetramers, (4) longitudinal assembly of ULFs to form immature IFs, which are $\sim$16 nm in diameter and finally, (5) radial compaction to form mature IFs $\sim$11 nm in diameter [89] (Fig. 1.18a).

Although the individual subunits of IFs are primarily α-helical, each has a head and tail domain as well as frequent interruptions, or linkers, in the coiled-coil backbone. Crystallisation of intermediate filaments has not yet been achieved, possibly due to their fibrous nature. Several structures of IFs have been partially solved by expressing

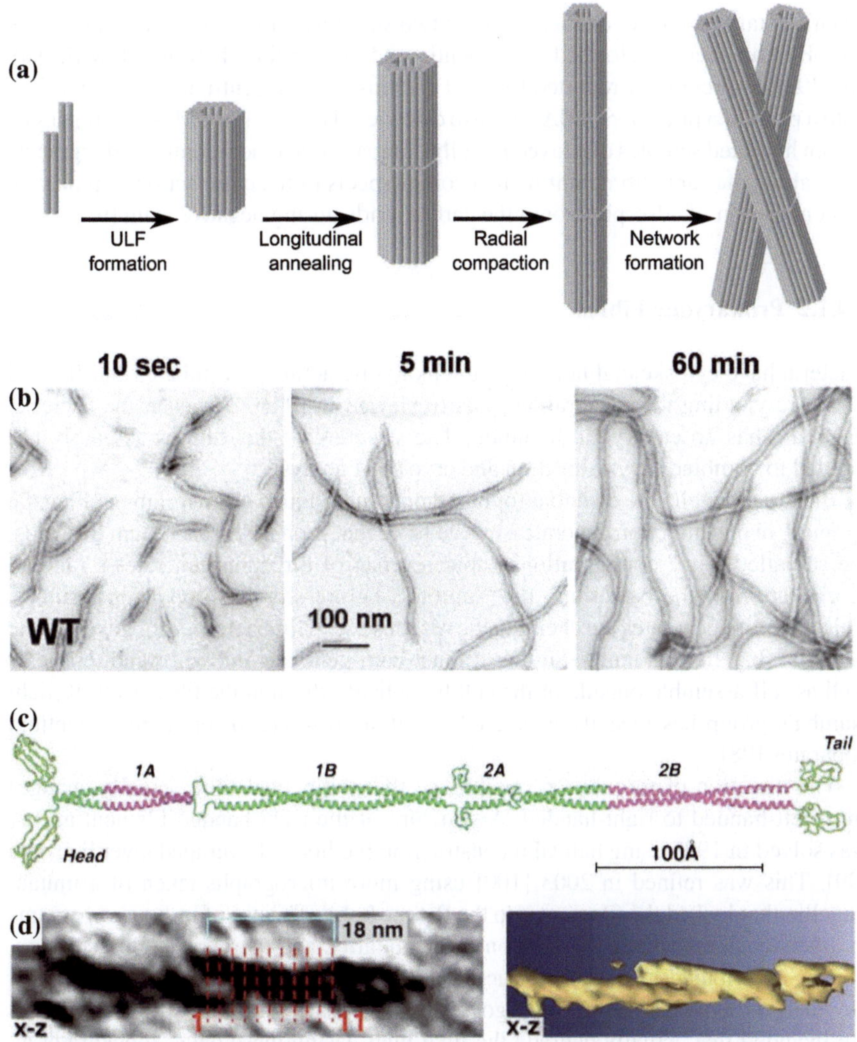

Fig. 1.18 IF assembly and morphology. **a** and **b** IF assembly occurs over 1 h and involves 5 steps. The interactions between IF subunits and filaments are unknown (modified from [90]. Copyright (2005) National Academy of Sciences, U.S.A.). **c** Crystal structure fragments and modelling of an intermediate filament dimeric coiled coil. Magenta regions have been crystallised, *green* are computationally modelled [91]. **d** Electron cryo-tomography of a frozen-hydrated IF at ~30 Å resolution. A right handed supercoiling is clearly visible, with a pitch of ~960 Å [92, 93]

fragments of coiled coil and crystallising them before either stitching the models together or simulating the adjoining protein [94] (Fig. 1.18c), and reviewed in [95]).

The filamentous structure can be imaged at low resolution by negative stain and cryoTEM [92]. CryoTEM immediately increased the understanding of IF structure by showing long, stiff regions separated by partially unwound "kinks", whereas negative stain had never delineated these two structures. Electron cryo-tomography (cryoET) has been performed on vimentin and shown that there is a low density core [92], as previously reported for hard keratins [96]. A right-handed supercoiling with a pitch of a pitch of ~960Å was also discovered (Fig. 1.18d). Most interestingly, frozen hydrated samples displayed radially compacted and non-compacted segments of IF along the same fibre, hinting that some aspects of the compaction step may be a phenomenon of adsorption onto the carbon grids during negative stain [93].

1.4.1.2 Prokaryotic Fibres

Bacteria have cytoskeletal homologs to eukaryotic actin, microtubules and IFs and these are yielding to structural analysis (reviewed in [97]). However, the bacterial flagellum has no eukaryotic homolog. The structure of this fibrous assembly has yielded to combined crystallisation and cryoTEM analysis.

Bacterial flagella are essential for movement in an aqueous environment. Flagella spinning in one direction generate a motile force that propels the bacterium forwards, the so called "run" configuration. Rapid reversal of direction generates a random reorientation of the bacterium, the "tumble", before straight propulsion resumes. This therefore generates the chemotactic response seen for bacteria with flagella. The bacterial flagellum is unique in that it must both generate and withstand torque as well as self assemble outside of the cell by delivery through the fibre itself. Keiichi Namba's group has recently generated an all-atom model of the entire flagellum apparatus [98].

The direction of torque is reversible, so that the helical fibre actually changes from left-handed to right-handed. A structure of the right-handed filament at 9 Å was solved in 1995 using helical reconstruction methods of averaged layer line data [99]. This was refined in 2003 [100] using more micrographs taken of a mutant flagellin that locked the filament into the R-type helix. The use of in-house programs to assess each micrograph so that only high-quality, and therefore high resolution, images contributed to the final structure [101]. This is in contrast to single particle analysis where selecting only good micrographs is difficult, so "bad" images are included that actually degrade the final map. Using these data, Yonekura et al. achieved atomic resolution detail from only 102 filaments by helical reconstruction [100]. Furthermore, this allowed an all-atom model of the subunit, flagellin, to be built. This was achieved by using the truncated crystal structure, solved by X-ray diffraction [102], as a guide to model the full protein in situ (Fig. 1.19). The resulting structure showed a very small channel 2 nm wide that exported flagellin must traverse, meaning that unfolding must precede export.

Recently, the group of Namba have solved the structure, by cryoTEM and helical reconstruction, of the left-handed filament [103]. This allows direct comparison of the two flagella structures in both the run and tumble conformations (Fig. 1.19). The result of this project is a detailed understanding of flagellum assembly at both the visual and biochemical levels which could not have been achieved without the use of cryoTEM and various reconstruction methods.

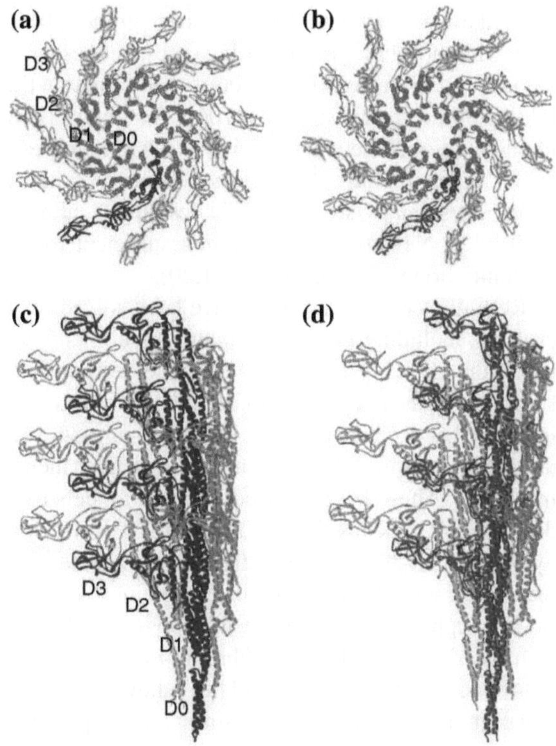

Fig. 1.19 The structure of R- and L-type flagellum solved by cryoTEM and helical reconstruction. **a** and **b** Top view of the L- and R-type flagellum. **c** and **d** Side view, showing the small tilt to the *left* and *right* of the L- and R-type flagella, respectively. Modified from [103]

1.4.2 Amyloid

Fibres based on the β-sheet folding motif are grouped under the generic term "amyloid", which was confirmed to have fibrillar content when it was first imaged by TEM in 1959 [104]. Many diverse proteins have a propensity to form amyloid-like assemblies; even those with a low β-sheet content can form predominantly β-sheet fibrils after denaturing [105]. However, ubiquitous as the fold appears to be there are relatively few structural details known about the fibres. This is partly due to structural heterogeneity; the fibre morphology changes depending on the protein used, which is compounded by the fact that within each preparation fibres show varied morphologies [106–108]. Due to these ambiguous results it has proven hard to elucidate the superstructure or functionalise amyloid fibres, although their structural properties have been exploited [109].

Despite these factors, several generic characteristics of amyloid fibres are known. Under X-ray fibre diffraction a cross-β pattern is visible, with a meridional reflection at 4.7 Å and an equatorial reflection at 10 Å. These correspond to the inter-β strand distance and inter-β sheet distance, respectively. Their orientation indicates that the β-strands are perpendicular to the long axis of the fibre and form sheets parallel to the fibre, like a ladder where the rungs are β-strands [105]. By TEM the fibre dimensions vary, with morphologies including tapes, helices and twisted ropes [108]. CryoTEM has allowed low resolution structure determination that indicates extensive un- or re-folding of the protein subunits [106]. The fibres also have a left-handed supercoiling in unidirectional metal shadowing and AFM studies [108, 110, 111].

3D reconstruction of the electron density is limited by the variable pitch of the fibres. Most maps are >25 Å in resolution, although a handful are below 15 Å [110, 111]. Nevertheless, this is not sufficient for anything more than building in randomly oriented β-sheets (Fig. 1.20).

Recently, crystal structures have been solved for fragments of amyloid-forming proteins, including insulin, β_2-microglobulin and amyloid-β for which 3D density maps already exist [112], as well as other amyloidogenic peptides [113]. These showed how the side chains interdigitated to form closely associated β-zippers with interiors that exclude all water. This also suggests a scheme of polymerisation where the zipper faces must first become exposed from their native fold before sequestration into fibres. These have allowed all-atom models to be built based on previous EM data, as well as X-ray fibre diffraction [113], which are remarkably similar to those hypothesised without informative crystal structures [105, 108] (Fig. 1.21), but with increased rationalisation and understanding.

Although most electron density maps derived from cryoTEM data show smoothly distributed density along the fibre, those derived from images of β_2-microglobulin [114] and prion protein (PrP) [115] show sub-domains along the fibre. These are more reminiscent of beads on a string and suggest that the rungs of a ladder model is incomplete, or perhaps not applicable in all amyloid-like systems. Indeed, models of β_2-microglobulin based either on cryoTEM data of fibres [114] or X-ray diffraction from non-fibre forming peptides [116] describe differing morphologies, although these models may be mutually compatible, since they are based on different data. They may also reflect the polymorphic nature of amyloid fibres, since even small fragments of amyloidogenic proteins show differing crystal packing and orientations [112].

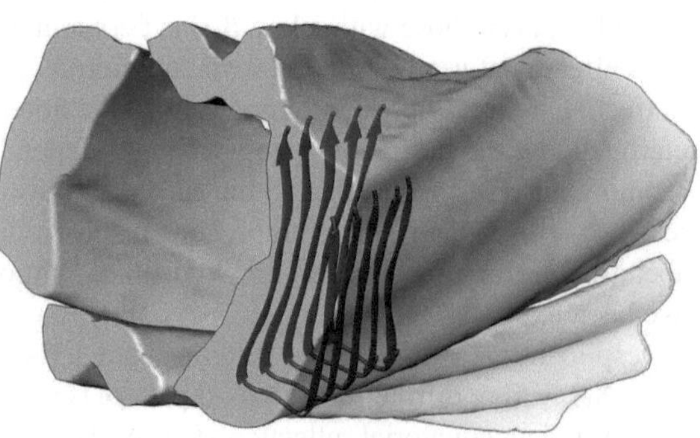

Fig. 1.20 Low resolution model of amyloid fit into a cryoTEM map. The backbone model (PDB code 2BEG) was fit with no specific orientation or interactions modeled into the EM map (EMDB code 5052) [110]

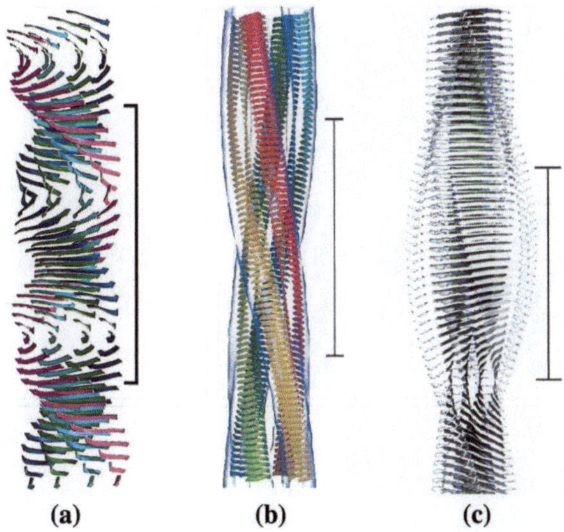

Fig. 1.21 Generic models of amyloid fibres. Models modified from [105] **a**, [108] (Copyright (2002) National Academy of Sciences, U.S.A.) **b**, and [113] **c**. Scale bars are all 10 nm

1.4.3 De Novo Designed Fibres

Interest in protein-based fibrous biomaterials has increased recently as they have potential applications in biotechnology and synthetic biology; for example, as scaffolds for 3D cell culture, tissue engineering and templating the assembly of functional inorganic materials [117–119]. Whilst natural proteins can and are being used in these areas, much simpler, or stripped-down, systems are preferable as they reduce complexity and potentially allow better understanding and control over the folding and assembly processes leading to fibre formation.

The past decade has witnessed a plethora of successful peptide-based designs of fibrous biomaterials encompassing peptide-organic hybrids [120], β-hairpins [121] and peptide amphiphiles [122], all of which are based primarily on β-sheet structures [119]. Recently, fibres based on protein fragments have been rationally designed [123, 124]. However, no design process is complete until the structures of the components and their assemblies have been determined at, or near to atomic resolution. In contrast to the natural fibres described above, the structures of which have been solved by combining X-ray diffraction data with electron density maps derived from cryoTEM, *de novo* fibrous systems are less well understood at the supramolecular level. EM has been instrumental in proving such systems form fibres and/or gels, and relating the rheological characteristic of the bulk material to the morphology of the superstructure [125]. However, although *in silico* models for several of the above systems have been presented (Fig. 1.22), there is currently no high-resolution structure for a fibrous biomaterial of *de novo* design.

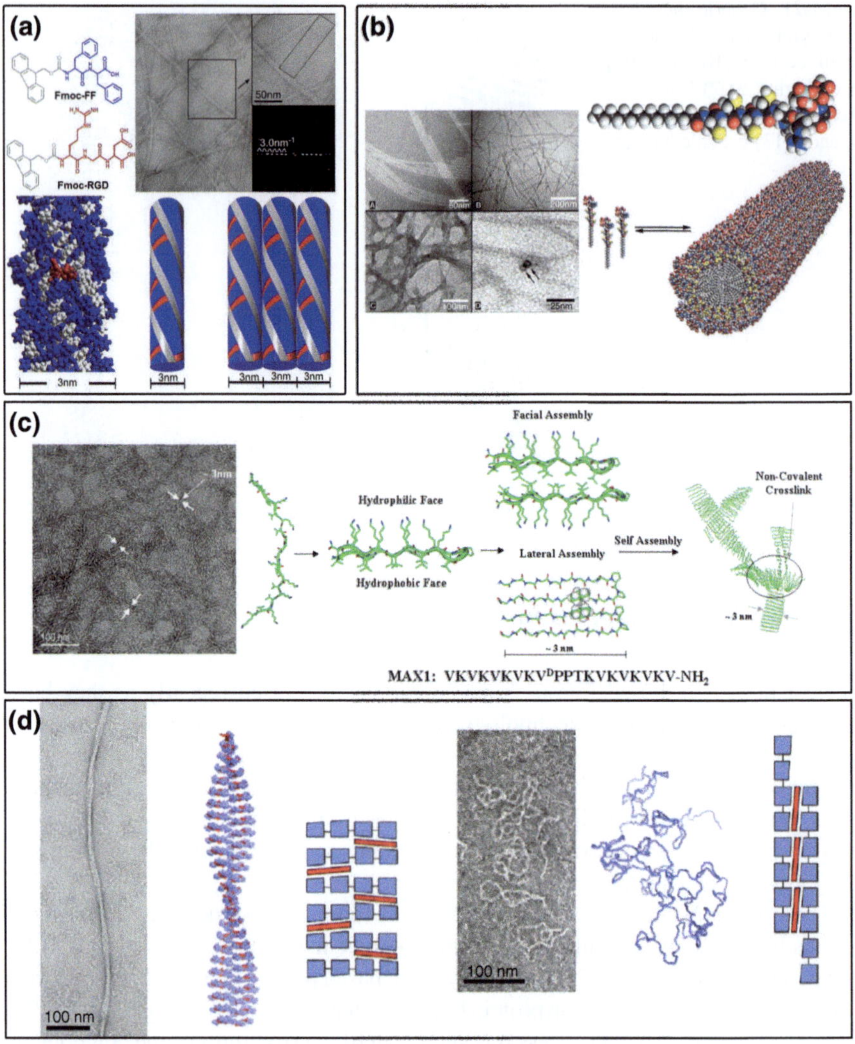

Fig. 1.22 *In silico* models of *de novo* designed fibres. For each design a respective EM image is shown along with an *in silico* model. **a** Fibres formed from Fmoc dipeptides held together with hydrogen bonds between the backbone (*blue*) and π-stacking between Fmoc groups (*grey*) (modified from [120]). **b** Tubular micelles formed from peptide amphiphiles (modified from [122]. Reprinted with permission from AAAS). **c** Inclusion of D-proline enforces a β-hairpin which assembles to form β-sheet fibres 3 nm across (reprinted (adapted) with permission from [126]. Copyright (2009) American Chemical Society). **d** Engineering the natural protein titin (*blue*) allows fibres of differing morphology to form after the addition of telethonin (*red*), a small titin crosslinking protein (reprinted (adapted) with permission from [123]. Copyright (2010) American Chemical Society)

1.4.4 Self-assembling Peptide Fibres

In the Woolfson group, α-helical coiled coils have been used to design fibrous biomaterials [127]. The sequence to structure rules governing coiled-coil formation are arguably the most well understood of all protein folds [128]. Coiled coils comprise two or more α-helices that can be either parallel or antiparallel and supercoil around one another with a left-handed pitch (Fig. 1.23a). The sequence of the individual α-helices has a distinctive pattern called the heptad repeat; *HPPHPPP*, where *H* is a hydrophobic and *P* a polar residue, and is assigned the register *abcdefg* (Fig. 1.23b). Although α-helices have ~3.6 residues per turn, the heptad repeat gives a repeating pattern of hydrophobic residues every 3.5 residues. Thus, the hydrophobic seam wraps around the α-helix with a left-handed twist. The α-helices form distinctive knobs-into-holes packing, where the hydrophobic amino acid "knobs" at *a* and *d* interdigitate with corresponding "holes" on the neighbouring α-helix. The heptad repeat can be illustrated as a helical wheel (Fig. 1.23c); the *a* and *d* positions form the hydrophobic core down the middle of the coiled coil (grey area). The combination of residues at *a* and *d* affect the oligomerisation state of the coiled coil; isoleucine and leucine at *a* and *d*, respectively, tend to give parallel dimeric coiled coils [129].

These rules governing coiled-coil formation can be distilled into design rules. Residues at the e and *g* positions can be given charge patterns that promote parallel dimer formation and provide additional stability by forming intermolecular salt bridges [128]. Dimers can be further selected for by placing an asparagine residue

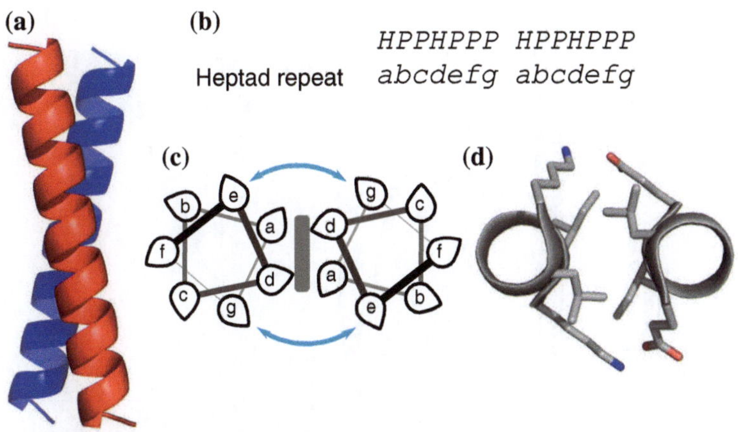

Fig. 1.23 Coiled coil design rules. **a** Cartoon representation of a dimeric coiled coil, α-helices are shown as *blue* and *red*. **b** Heptad repeat, *H* and *P* are hydrophobic and polar residues, respectively. **c** Helical wheel schematic illustrating amino acid positions. The view is from the *top* of the coiled coil down through the center. Thickness of *lines* increase with the height of the residues in the α helices and the pointed ends represent the Cα–Cβ bond direction. The hydrophobic core is shown as a *grey box*, whereas electrostatic interactions are shown as *blue arrows*. **d** Representative crystal structure of a coiled coil showing the hydrophobic core and stabilising Lys–Glu salt bridges

at one of the *a* positions, which forms complementary hydrogen bonds within the hydrophobic core. The remaining residues can be used for specific roles. For example, the *f* position is frequently used for adding a chromophore to aid characterisation by UV-vis spectrometry and analytical ultracentrifugation, or the *b*, *c* and *f* positions can be used to promote helicity by including alanine residues, or unnatural amino acids for derivatisation and functionalisation.

These design rules were exploited to construct a *de novo* designed self-assembling fibre (SAF) system [130]. In these so-called SAFs, two complementary 28-residue peptides combine to form a sticky ended dimeric coiled coil building block of dimensions ~2 × 6 nm (Fig. 1.24). These assemble to form stiff fibres on average 84 nm wide and 42 μm long [131]. Interestingly, after negative stain the SAFs display lateral striations along the entire length and width of the fibre (Fig. 1.24c). These have a periodicity of approximately 4.2 nm, which is the approximate length of a 28-residue peptide when folded into a coiled coil (28×0.148 nm $= 4.14$ nm, where 0.148 nm is the rise per residue of a coiled coil). The striations are present after negative stain with either cationic or anionic stains, and when both are applied. They are also visible after washing, indicating that they are in fact acting as a *positive* stain and interacting directly with the sample, not simply as a topographic indicator (as negative stain ideally works). In an attempt to elucidate what is generating these striations the candidate moieties present on the SAFs were mutated. Capping the N- or C-termini by acetylation or amidation, respectively, fails to negate striation formation [132]. Indeed, a mixture of both capped peptides does not form fibres, indicating the termini are very close in space within the SAF. The striations also change periodicity if the SAF peptides are lengthened, indicating a sequence-dependence [132, 133].

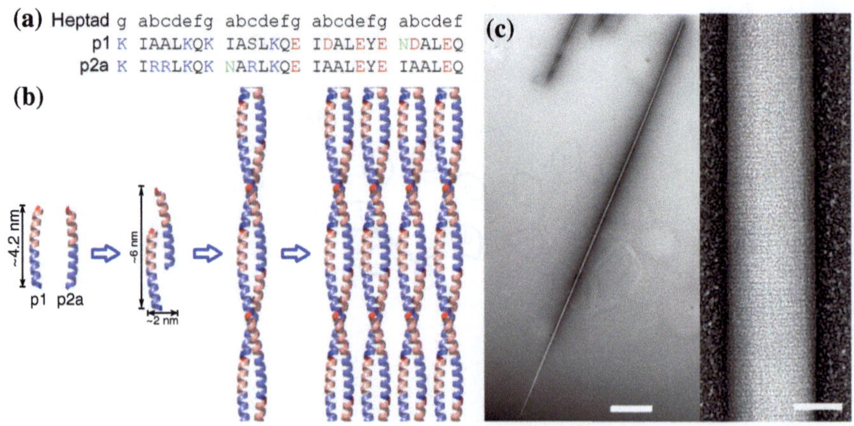

Fig. 1.24 Design of a sticky-ended coiled coil system. **a** and **b** Two peptides (p1 & p2a) assemble to form a heterodimeric sticky ended coiled coil that associates further to form hexagonally packed fibrils. Heptads are coloured for overall charge; *blue* and *pink* for basic and acidic, respectively. N- and C-termini are shown *darker blue* and *red*, respectively. **c** Fibrils associate laterally to form fibres which display striations after staining with heavy-metal salts and visualisation by TEM. The scale *bars* represents 2 μm (*left*) or 50 nm (*right*)

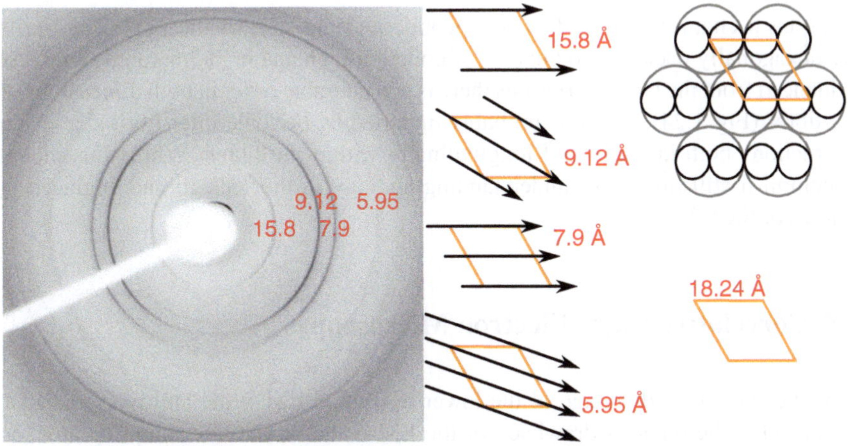

Fig. 1.25 Wide-angle X-ray diffraction pattern of partially-aligned SAFs [132]. The fibre orientation is vertical. Equatorial reflections (in Å) describe the packing regime of the coiled coils within the SAFs. A pictorial explanation of these reflections is also provided. *Black circles* represent the α-helices and *grey circles* the path they trace within the coiled coil. The orange parallelogram is the unit cell, with *black arrows* the lattice planes described. The view is along the fibre long axis

Although the reason for lateral thickening has remained a mystery the mode of thickening has been discovered. Wide-angle X-ray diffraction (WAX) showed a series of equatorial reflections that correspond to hexagonally-packed coiled coils (Fig. 1.25) [132], although the reason for this thickening could not be deduced from the experimental data.

Recently, the full fibre formation pathway has been determined by a combined peptide design, biophysical and microscopical approach [134]. The SAFs polymerise *via* a process of nucleation. The nucleating species comprises ~7 heterodimers, from which the fibre grows both laterally and longitudinally until a period when growth ceases to be in 3 dimensions and instead becomes epitaxial (Fig. 1.26).

A low resolution model (~20 Å) for the structure based on negative-stain TEM and X-ray analysis of dried and partially aligned fibres has been proposed [132]. The sticky ended dimers assemble end-to-end into α-helical fibrils. These fibrils are most likely extended dimeric coiled coils, albeit interrupted every 28 residues

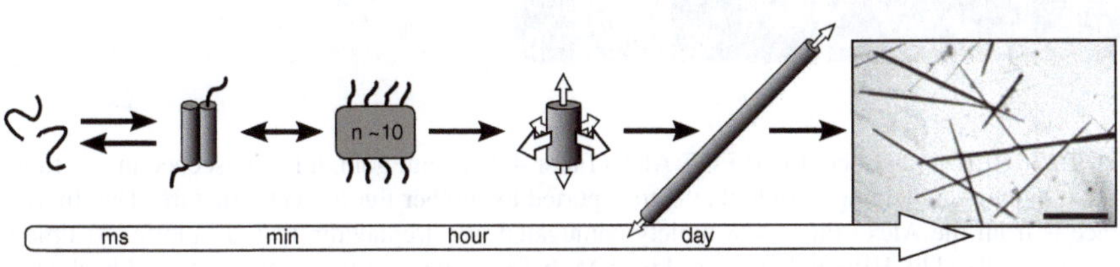

Fig. 1.26 Experimentally-verified assembly pathway of the SAFs (modified from [134]). Scale bar is 10 μm

along each chain by the end of one peptide and the start of the next. The fibrils are hexagonally packed into bundles of coiled-coil chains in the matured fibres. As evidenced from the lateral striations there is considerable order in both dimensions of the fibres (Fig. 1.24), which is probably imparted by periodic inter-fibril side-chain interactions both laterally and longitudinally within the fibres. Whilst this model is helpful it still limits our understanding and any further design and engineering studies of the SAFs.

1.5 Correlative Light Electron Microscopy

Whereas the clear advantage of the lower resolution LM is the ability to visualise whole cells, the obvious drawback of the EM is the diminutive sampling area; as reported in 2009, only around 10^3 mm^3 of material has been imaged since the commercial availability of the TEM [34]. Combining the advantages of LM and EM is highly desirable, as it offers real-time imaging of the live cell with the high spatial resolving power of the EM. Correlative light electron microscopy (CLEM) has been a viable technique for over two decades [46], although interest in CLEM has increased in recent years due to the widespread use of AFPs and advances in LM.

Correlating live cellular events with their ultrastructural effects is the primary objective of CLEM (Fig. 1.27). The obstacles facing such a task are many [45]. Fixing the cell in the same state as seen in the LM is the initial aim. Next, the same location, cell and organelle/region need to be traced back. Finally, the same fluorescent probe imaged in the LM must be identified in the EM, which means it must also be, or produce something that is, electron dense. HPF is the ideal method for rapid fixation of samples, with a time-resolution of 4 s [46], although chemical

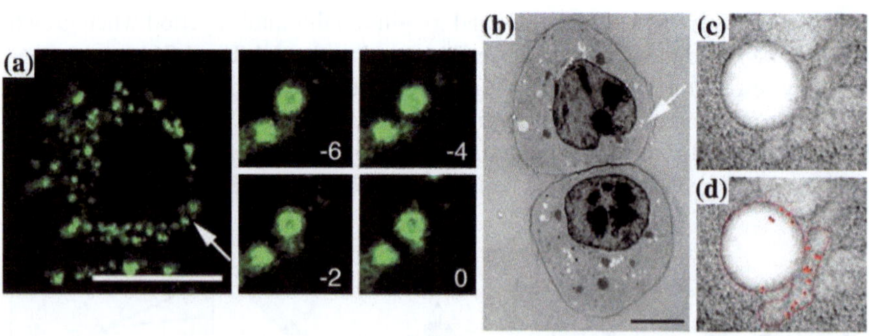

Fig. 1.27 CLEM of endocytosed EGF-Alexa Fluor 488-10 nm gold. **a** Fluorescence image time-course showing a budding vesicle that is transported to another fluorescent structure. The fluorescence is from the Alexa Fluor 488 moiety conjugated *via* streptavidin to EGF-biotin. At time 0 the cell was fixed by HPF and processed for EM. **b** TEM image of the same cell traced back from LM. **c** The high magnification image clearly shows a spherical vesicle that has joined to a tubular structure. **d** The 10 nm gold is now visible as dense colloidal particles. Modified from [46]

fixation of cell monolayers is also extremely fast (<1 min) [135]. Tracing back the same cell can also be achieved either with a coverslip with a finder grid etched into it that cells can grow on, or locator grids that sit on top of cells [46]. These are visible in both the LM and during processing for EM, so the same cell can be located post-fixation and sectioning.

The real challenge of CLEM lies in locating the same structure seen in the LM. This can be achieved in one of two ways; either whole FLM and EM images can be directly correlated, such as for fixed thin sections dual-stained for both LM and EM, or by correlating the same probe seen in both modalities. For the former, correlated light electron microscopes have recently started to become available. For the latter, this requires a probe that is both fluorescent and electron dense (Fig. 1.27). Superresolution FLM will aid the field of CLEM enormously, since the use of AFPs enables live cell imaging of proteins *in vivo*. However, so far superresolution is a relatively slow technique; PALM has a temporal resolution of ~30 s and cannot be combined with optical sectioning (reviewed in [136]). Whereas STED has a temporal resolution of ~35 ms, the spatial resolution is "only" 62 nm, larger than most vesicles [18], and fluorescence quenching rapidly becomes limiting [137]. With the ultimate aim of live cell CLEM superresolution techniques will need to improve but, once they have, should be the perfect counterpart to CLEM (Fig. 1.6).

1.5.1 Correlated Light Electron Microscopes

With an increasing interest in CLEM has come the development of equipment, including the production of an integrated light and electron microscope (ILEM) [138]. This has a FLM inside the column and perpendicular to the electron beam, so the sample must be rotated between modes (Fig. 1.28). Correlation between the two modes is

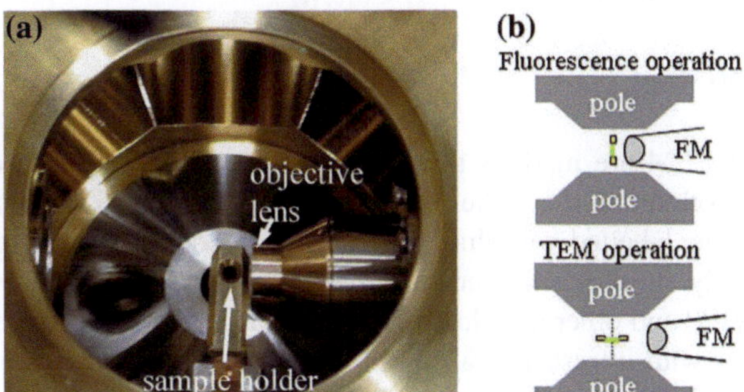

Fig. 1.28 Integrated light and electron microscope. **a** Photo of the LM objective lens and sample holder within the TEM column. **b** Schematic showing how the mount rotates between FLM and TEM modes. Modified from [138]

simplified by computationally calculating the area of interest in the LM for recognition in the TEM mode. For ILEM the sample must be static and thin for TEM compatibility, which precludes live cell CLEM, although this can be circumvented somewhat by imaging just prior to fixation. The main advantage lies in the ease of area of interest localisation, which can be compared directly by both LM and EM.

1.5.2 CryoCLEM

EM of biological samples suffers from the uncertainty that what is being imaged reflects the true nature of the sample. As discussed above, chemical fixation can distort the sample, so for increased fidelity biological specimens can be rapidly frozen, either by HPF followed by CEMOVIS for large samples, or liquid ethane plunge freezing for thin samples.

Following vitrification, finding the area of interest is necessary. Due to the low contrast of biological specimens and the damage caused by electrons, the area must be located quickly, reliably and with the least electron dose possible [51]. FLM is the ideal technique with the high sensitivity and large field of view, and several designs of cryo-FLM stages have recently been developed [139–141]. Recently, ILEM has been complemented by cryo-ILEM, a technique whereby frozen-hydrated samples or sections can be viewed in a near-native state by both FLM and EM. This was achieved by simply replacing the conventional grid holder with a cryo-holder [142]. The advantage of such systems are a direct correlation between the sample in the LM and EM, although live cell CLEM is again non-trivial with such a workflow. Indeed, combining live cell imaging with rapid plunge freezing can alter the conformation of the areas of interest [143].

1.5.3 CLEM Markers

1.5.3.1 Affinity Tags

Essential for CLEM are markers that are both fluorescent and electron dense for imaging with both modalities. The first experiments were performed on serial sections for EM, one labelled with fluorescent antibody conjugates and the other with gold, a relatively simple technique still common today [144]. However, with the discovery of GFP and other autofluorescent proteins (AFPs), the desire to correlate live cell events with ultrastructural information has come to the fore. To achieve this, various markers and methods are currently available, each with advantages and disadvantages. Traditional methods, such as antibody labelling with colloidal gold and photo-induced precipitation of osmiophilic diaminobenzidine (DAB) are either invasive or destructive. More recent methods, including using quantum dots (QDs) or the FlAsH/ReAsH biarsenicals, are restricted by membrane permeability or reduced resolution.

Antibodies to fluorescent probes conjugated to colloidal gold can be used as CLEM probes either pre- or post-fixation. Usually antibodies are incubated with samples post-fixation that have either been permeabilised with detergent or processed using the Tokuyasu method [145]. This is because antibodies are very large, approximately 150 kDa, and when conjugated to colloidal gold (5–10 nm) become even larger (so-called immunogold labels). The permeability of fixed samples is very low, so without permeabilisation antibodies label only the surface of samples [146]. Antibody labelling is also a relatively low resolution technique. IgG is approximately 15 nm wide, and when used as a secondary antibody conjugated to colloidal gold, the gold can appear up to 25 nm away from the target protein. This is reduced with the use of protein-A conjugated to gold (PAG). PA binds to the Fc region of most species of IgG, but the gold can still appear up to 15 nm away from the protein [147].

Immunolabelling intracellular structures for EM (immunoEM) is hampered by GA fixation, which destroys antigenicity. Ultrastructure is also deformed during the dehydration and embedding steps. To circumvent this, Tokuyasu developed a technique whereby cells can be frozen and sectioned after embedding in sucrose, which retains ultrastructure and avoids dehydration and plastic resins [145]. Light fixation with PFA also helps retain antigenicity, and the thawed cryosections have greater permeability. These factors contributed to making the Tokuyasu method ideal for immunoEM, which has led to its use in CLEM [147]. ImmunoEM following HPF has also recently become a viable option [148]. It is difficult to perform live-cell CLEM using the Tokuyasu technique due to sample preparation. Generally, the specimen is formed into a pellet after fixation prior to sectioning, which means it is impossible to track one cell or maintain morphology. Nonetheless, static single-cell CLEM can be performed by flat-embedding procedure, which retains the in situ orientation of the sample. This can be followed with fluorescent antibodies imaged by FLM to find regions of interest before labelling with gold-conjugated antibodies and imaging the same regions at high resolution [149]. This technically-demanding method has recently shown promise with live-cell CLEM using GFP-expressing neuronal cells [150] (Fig. 1.29), although it has not yet found widespread use.

Alternatively, antibodies to endocytic proteins, or endocytic proteins themselves, may be conjugated to both a fluorescent probe and colloidal gold to form a single bifunctional probe and incubated with live cells (Fig. 1.27). The bifunctional probe is then endocytosed and segregated in the endocytic pathway [46]. This allows live cell imaging of the fluorescent probe (usually an Alexa dye) prior to fixation and tracing back the same probe in the EM which is now visible as 5 nm or 10 nm colloidal gold (Fig. 1.27). The drawback to this method is that only endocytic events may be studied.

A new bifunctional probe called fluoronanogold (FNG) is also based on affinity labelling, either *via* antigenicity or streptavidin-biotin binding [146]. Since the antigenicity comes from only the fragment of antigen binding (Fab′), and the colloidal gold particle is replaced by a 1.4 nm gold cluster, this is much smaller than antibody-conjugated probes. Since the gold cluster is a molecule, not a metallic particle,

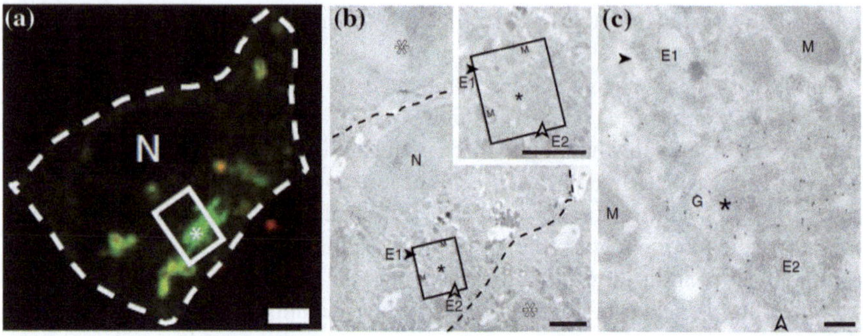

Fig. 1.29 CLEM performed using the Tokuyasu technique. **a** Optical section of a cell expressing LAMP-1-GFP (*green*) after fixation. **b** and **c** ImmunoEM of ultrathin section corresponding to the optical section. IgG is bound to GFP and visible *via* PAG. N, G and M denote nucleus, Golgi, and mitochondria, respectively. E1 and E2 are endosomes, Modified from [150]

monofunctionalisation is possible, meaning that single particles can be localised to single proteins [151]. This is not possible with colloidal gold suspensions, where each particle is free to adsorb more than one antibody molecule. However, the question of whether the nanogold and fluorescent moiety were both linked to the same probe was not addressed, meaning that the a fluorescent signal may not represent an electron dense particle in the EM. Also, the 1.4 nm nanogold is too small to be seen directly unless visualised by STEM [151]. Alternatively, autometallography by silver or gold-enhancement can be used to develop the probes to sizes visible by conventional TEM [152].

Similarly, QDs can be used to study endocytic events. QDs are brightly fluorescent, semiconducting nanocrystals, usually composed of either CdSe or CdTe [154], which form a core that is electron dense enough to be visible in the EM [153]. QDs can be conjugated to antibodies or other targeting molecules, such as protein-A or streptavidin [155]. Because the colour of the QD is related to their size [156] it is possible to discriminate between at least 3 different QDs under both LM and EM based on size and shape; spherical QD525 (blue) and 565 (green) and prolate QD655 (red) (Fig. 1.30) [153]. However, although cadmium is more electron dense than biological specimens, it is less dense than contrasting agents used, such as uranyl acetate, which can mask the QDs [157]. As a probe for CLEM, QDs suffer from the same drawbacks as other exogenous labels; their large size limits sample penetration in fixed cells, and thus far QDs are limited in their use to endocytic events or post-fixation studies.

Aside from the low permeability of affinity tags, the population of proteins that are labelled are only a subset of the whole population. This is the result of a trade-off between specific and background labelling requiring a less-than 1:1 ratio of marker to target protein. The ideal scenario is with fusion proteins, where each protein is genetically tagged with a visible label.

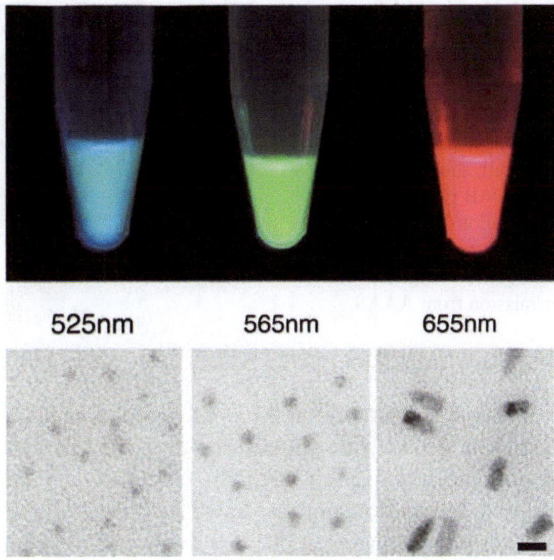

Fig. 1.30 Quantum *dots* suitable for CLEM. Scale bar is 10 nm. Modified from [153]

1.5.3.2 Genetically Encoded Tags

Genetically encoded reporter proteins are the most reliable tags for protein targeting; the resulting fusion proteins result in 1:1 stoichiometry of target protein to marker, with GFP as the archetype. However, concerns arose over the size of AFPs (27 kDa), which are often larger than the target protein. Fluorescein bi-Arsenical Hairpin-binding (FlAsH), based on highly specific metal chelation of Arsenic (III) to cysteine thiol pairs, were developed in 1998 [158]. The arsenic ions bind a CCxxCC motif (where C is cysteine and x is any amino acid, known as a 4C-tag). ReAsH, a Resorufin-derived red form of FlAsH, was developed in 2002 for multicolour tagging [159]. FlAsH and ReAsH are suitable for live-cell experiments; both are membrane permeable and highly penetrating. They are also non-fluorescent until bound to protein, reducing false-positive staining [160]. Although not intrinsically visible in the EM, ReAsH was found to photo-polymerise DAB to an osmiophilic precipitate that is visible in the EM and hence was compatible with CLEM [159]. Several studies have demonstrated the value of biarsenicals with respect to CLEM. In one of the earliest studies in 2002, Gaietta et al. were able to visualise ReAsH-labelled connexin43 in a dual protein-age/ultrastructure study of connexin dynamics (Fig. 1.31) [161]. More recently, Lanman et al. used ReAsH in a CLEM study of Flock House virus location and assembly which would have been impossible with GFP due to steric hindrance [162]. Despite these advantages, biarsenicals have several limiting drawbacks. FlAsH is not able to polymerise DAB and is therefore not compatible with CLEM. Polymerised DAB can also drift from the site of production and will fill the organelle encapsulating it [163], meaning that CLEM performed with ReAsH is a low resolution technique.

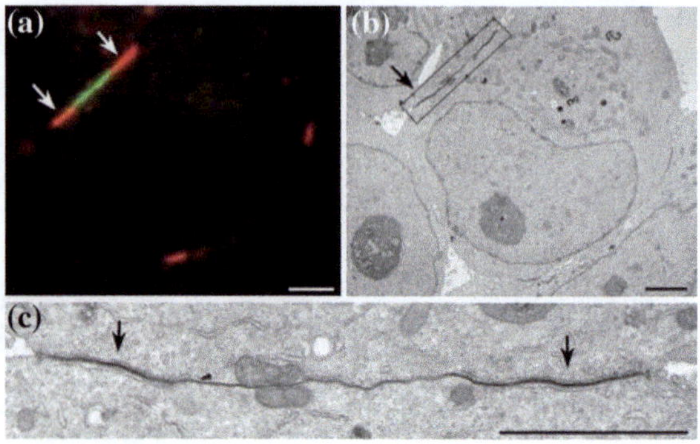

Fig. 1.31 CLEM of connexin43 by photoconversion of DAB with ReAsH. **a** Confocal image of FlAsH (*green*) and ReAsH (*red*) bound connexin43. **b** and **c** EM of the same cell, showing electron-dense DAB precipitate at the same regions as ReAsH was found in the LM. Modified from [161]. Reprinted with permission from AAAS

The major limitation of AFPs as probes for CLEM is the total lack of visibility in the EM; proteins are not electron dense and fluorescence is completely quenched by conventional fixation methods. Commercially available antibodies to GFP are available but suffer from the drawbacks listed above. One method of visualisation involves photoconversion of DAB by production of singlet oxygen ($^1O_2$) from fluorescing GFP [164]. Known as GFP-recognition after photobleaching (GRAB) the disadvantages of using DAB remain, although electron tomography has been demonstrated [165, 166]. A further drawback of GRAB is the low phototoxicity of GFP; although an advantage for LM, low phototoxicity is due to the negligible formation of $^1O_2$. This problem was circumvented by the labs of Ellisman and Tsien by genetically co-localising GFP with ReAsH, which behave as a FRET pair. Excitation of GFP could then be used to photoconvert DAB *via* FRET to ReAsH [163].

Tsien's lab has done much to improve the availability of high-resolution biomolecular imaging. They have produced AFP variants [30] (and see Sect. 1.2.1), developed ReAsH to be compatible with CLEM [161], designed a *de novo* protein that fluoresces infrared for whole animal *in vivo* studies [167], and most recently designed a fluorescent protein (mini Singlet Oxygen Generator; miniSOG) that produces a high concentration of oxygen radicals suitable for efficient DAB polymerisation and localisation in the EM [168]. MiniSOG is a mutated domain of phototropin, which binds with high affinity to flavin mononucleotide (FMN) and has a molecular weight of 15.3 kDa, nearly half the weight of GFP. After mutagenesis, miniSOG was fluorescent and could photoconvert DAB with high efficiency (Fig. 1.32). Localisation with numerous proteins proved CLEM was possible, even with the low resolution of

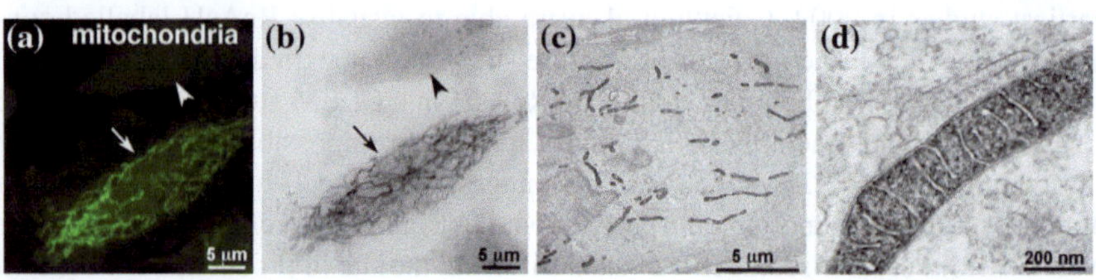

Fig. 1.32 CLEM of mitochondria using miniSOG. Images are, from *left* to *right*; confocal fluorescence (**a**), transmitted light (**b**), and electron microscopic (**c**, **d**). Modified from [168]

DAB. However, if simply identifying the same area as in the LM is the aim of the experiment, DAB precipitation with miniSOG should prove to be a viable technique.

None of the above techniques produce an *in vivo* electron-dense substance in live cells. Instead, either antibody and streptavidin conjugates must be added to fixed cells or limited to the endocytic pathway, or molecules (ReAsH and DAB) must be added to live or fixed cells and then processed to form electron dense materials. These limitations have been overcome for the light microscope with the discovery of GFP (described in Sect. 1.2.1), a genetically-encoded autofluorescent protein. An analogous probe for EM, one that is "auto-electron dense", would greatly increase the availability and pertinence of CLEM. In an effort to design such a probe, Wang et al. conjugated ferritin, a ubiquitous iron-binding protein, to GFP [169]. These produced iron-loaded nanoparticles clearly visible by cryoTEM and were applicable for electron cryo-tomography. However, the main drawback is that every ferritin subunit is attached to the protein of interest, which means that in Wang et al. there are 24 protein molecules attached to each particle visible in the TEM [169].

1.5.3.3 Overview

Each of the CLEM probes and techniques described above have their own advantages and disadvantages which were outlined in the relevant section. However, a few points should be highlighted here, and comparison of the size of the probes and their applications discussed here is shown in Fig. 1.33 and Table 1.1. Despite being established techniques, affinity-probe labelling (antibodies, Fab′, streptavidin) immediately increases the size of the probe to several 10's of nanometres. Even though it is genetically-encoded, ferritin is a large cage comprising 24 subunits. The newer probes ReAsH and miniSOG are much smaller, and this is highly advantageous. With affinity probes the final particle may not even lie within the bounding membrane [135], whereas with small, genetically-encoded probes the position of the particle will reflect the correct location of the protein of interest.

The ultimate goal of CLEM is a labelling strategy akin to that of GFP for LM, where a genetically encoded protein—that does not interfere with the endogenous function of the target protein—is readily and reliably visible by both LM and EM. Recently, focus has turned to superresolution and cryoEM, which are perfectly suited to visualising such probes. Indeed, superresolution is preferable to immunoEM since the labelling efficiency is orders of magnitude higher [9]. The only way to achieve this is by designing genetically-encoded probes that are visible by LM in a live cell and can be retraced in the EM.

1.5.4 Novel Probes for CLEM

GFP is an ideal label for LM because it is genetically encoded and auto-fluorescent in orthogonal cell types. An analogous tag for the EM would negate the obstacles

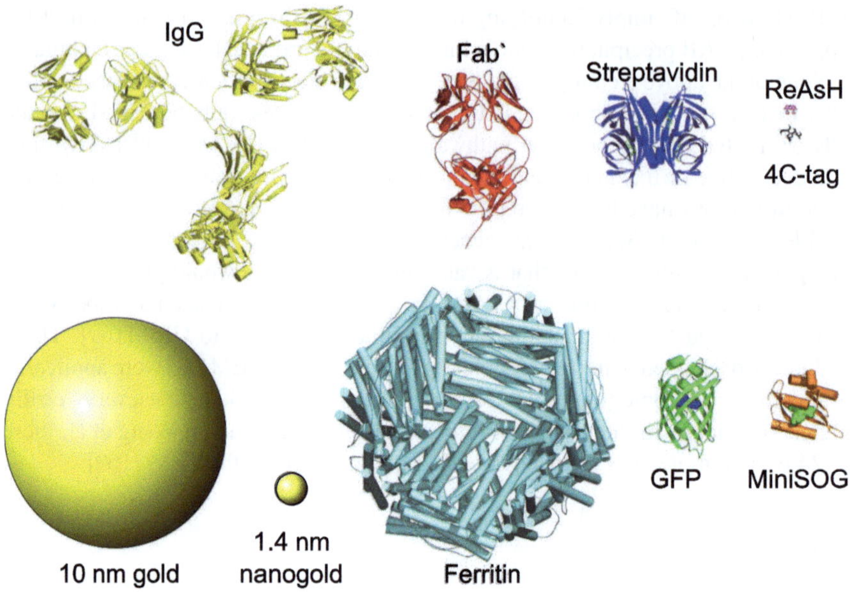

Fig. 1.33 Size comparison of probes currently used for CLEM. Each probe is to scale. Affinity tags and probes are shown in the upper layer. IgG (PDB code 1IGT) is not actually used as a probe but is conjugated to various particles that are, including colloidal gold and QDs. Fab′ is the fragment of antigen binding of IgG. Streptavidin and biotin (PDB code 1NDJ) can both be covalently linked to proteins of interest. Biotin is shown as *green* sticks. ReAsH binds to the tetracysteine (4C) tag. Colloidal 10 nm gold is approximately the same size as QDs. Nanogold is not a metallic colloid but in fact a defined single molecule. Ferritin is a homomer of 24 ferritin subunits (PDB code 1EUM). GFP is $\sim 3 \times 5$ nm in dimension, the chromophore is shown as blue spheres (PDB code 1EMA). MiniSOG binds to FMN (*green spheres*) and is approximately $^2/_3$ the size of GFP (PDB code 2Z6D)

associated with CLEM. Possible approaches include subcloning adjacent to GFP either; a heavy-metal binding protein, a protein that can be specifically tagged with an electron-dense particle, or a protein which can produce electron density.

For the latter, proteins that produce electron density are already available and described above. Photoprecipitation of DAB may be induced, for example with miniSOG [168]. However all of these techniques rely on exogenous DAB, they produce a non-particulate diffuse electron density, and as such are relatively low-resolution They are also not compatible with cryoCLEM or HPF due to the requirement of a laser-induced photoconversion step and a liquid milieu to convert the DAB.

With regards to the second option, a protein that can be specifically tagged, a mutant of FK506 binding protein 12 kDa (FKBP) with a mutation of Phenylalanine36–valine (FKBP12(F36V)) is an ideal candidate since it is small (12 kDa) with high affinity (94 pM) and specificity for a synthetic molecule [170]. However, labelling any intracellular protein requires transmembrane delivery of the

Table 1.1 Summary of various techniques and probes used for CLEM. Table modified from [147]

	AFPs (GRAB)	FlAsH/ReAsH	QDs	FNG	Immunogold	MiniSOG	Ferritin
Live-cell LM	+++	++	+[a]	+[a]	–	++	++
EM	–	–	+	+[b]	++	–	+
CLEM	–/+[c]	–/+[c]	+	+	+	+[c]	+
CryoCLEM	–	–	+	–	–	–	+
Superresolution [136]	+	-	+	+	–	NT	+
Membrane penetration in live cell	NA	+	–	–	–	NA	NA
Penetration after fixation	NA	NA	+	++	+	NA	NA
Multiprotein labelling for CLEM	–	–	+	–	+	–	–
Endogenous target protein	–	–	+	+	+	–	–[d]
Pulse-chase	+[e]	++	–/+[a]	–/+[a]	–	–	–
Specificity	++	++	+	+	+	++	++
Key references	[165, 166]	[161]	[153]	[152]	[149, 150]	[168]	[169]

Different techniques are scored from – (not suitable) to +++ (highly suited). "Membrane penetration" indicates the ability of the probe to cross the plasma membrane non-invasively to label the intracellular target. "Penetration after fixation" refers to the depth of labelling in the sample that the probe can achieve after fixation with GA; for GRAB, ReAsH and miniSOG DAB is the penetrating reagent. "Multiprotein labelling for CLEM" relies on different colors and sizes of the probe being distinguished in both LM and EM; for example, while there are two colors of FlAsH/ReAsH, only ReAsH photopolymerises DAB. "Endogenous target proteins" are those with no genetic modification. "Pulse-chase" is a live-cell imaging technique that allows the researcher to image the same protein at different time-points in different colors. "Specificity" is a measurement of the fidelity of labelling; AFPs label one to one due to the nature of genetic covalent modifications, whereas antibodies can label one protein more than once or not at all; FlAsH/ReAsH are comparable to antibodies. NT = not tested. NA = not applicable. [a] Specifically for endocytosis events or direct microinjection of probes. [b] Requires autometallographic enhancement or visualisation with STEM. [c] Requires photo-polymerisation of DAB and osmium staining. [d] Obligate oligomerisation. [e] Only for photoconvertible AFPs, such as Kaede (often membrane-impermeable) probe. A new system for cytoplasmic delivery has recently been invented and so this approach will be further described in Sect. 5.5.

The subcloning of a heavy-metal binding protein would be most advantageous since the approach is closely analogous to GFP, requires intracellular delivery of only small molecules or salts, and should function in any host cell. Although ferritin fulfils these criteria [169], the large size (Fig. 1.33) and oligomerisation of the probe limit this approach to the tagging of only cytoplasmic proteins. Furthermore, the obligate oligomerisation may interfere with normal cellular function. One other such protein that has the potential to form electron-dense particles visible in the EM is metallothionein.

1.5.4.1 Metallothionein

Metallothionein (MT) is a small, cysteine-rich metal binding protein; one third of the 61 amino acids are cysteine. Highly conserved throughout the eukaryotes and present in some prokaryotes [171], MT is a heavy-metal ion chelating protein involved in detoxification, homeostasis and metal-donation [172], as well as pathological roles in diseases such as Alzheimers [173]. Although unstructured in the absence of metal, binding of heavy-metal ions causes MT to form a bilobed shape separated by a short flexible hinge [171]. Each lobe makes up a nearly spherical 1.5–2 nm domain wrapped around the cysteine residues (Fig. 1.34). The C-terminal α-domain possesses 11 cysteine residues, whereas the β-domain contains 9. The α- and β-domains bind, respectively, 4 and 3 bivalent ions, including Cd^{2+}, Pb^{2+} and Zn^{2+}, whereas monovalent ions, such as Ag^{+} and Cu^{+}, bind with a higher stoichiometry of 6 per domain [174].

The potential for MT to be used as an *in vitro* clonable EM label has previously been demonstrated [175–180]. Mercogliano et al. designed a tandem dimer of MT domains (MT2) fused to maltose binding-protein and used Aurothiomalate (AuTM) as a gold-ion donor to demonstrate that concatenation of MT could form electron dense nanoparticles [176]. Nishino et al. built on this discovery by fusing a tandem

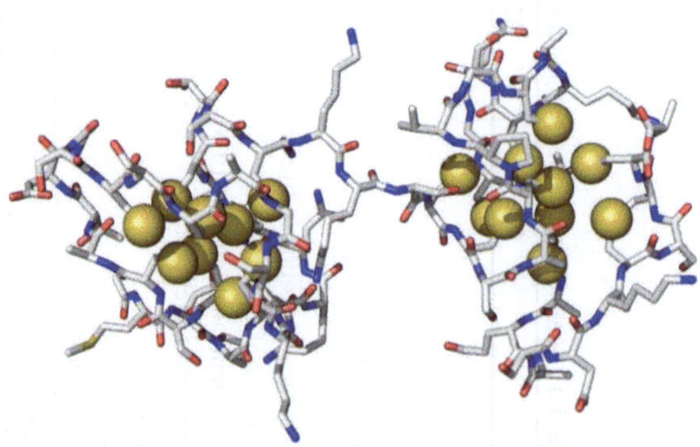

Fig. 1.34 The crystal structure of MT-2 (PBD code 4MT2), showing the thiolate core. Sulphur atoms of the cysteine residues are shown as *yellow* spheres. The α-domain is on the *right*

trimer of MT domains to the chaperonin GroEL, and illustrated that MT could be fused to oligomeric cellular proteins, without affecting assembly, and imaged in the TEM [177]. Instead of gold, Nishino et al. used cadmium chloride ($CdCl_2$) as the heavy-metal donor. There are so far only three reported examples of MT being used as an *in vivo* tag, and only one of these is in eukaryotic cells. Fukunaga et al. used MT3 to genetically label postsynaptic density-95 (PSD-95), a protein localised to the synaptic junction [178]. They report that oligomerisation of MT is required for visualisation, hence the use of PSD-95, which clusters at the synaptic cleft. Although the function of PSD-95 was unaffected, the cytotoxicity of $CdCl_2$ lowered the viability of the cells and disrupted the morphology. Diestra et al. localised four MT- or MT3-fusion proteins in *E. coli*; MT-RecA, MT-AmiC, MT3-MBP_{cyto}, and MT-Hfq [179, 180]. When treated with 10mM AuCl, numerous electron-dense particles were imaged at the expected locations which was confirmed by immunolabelling. However, AuCl is insoluble in aqueous media [181, 182], which raises questions about what the particles are formed from. Apart from $CdCl_2$, which is toxic, the only other heavy-metal compounds tested by Fukunaga et al. were mercury chloride ($HgCl_2$) and silver nitrate ($AgNO_3$), neither of which apparently bound MT3. It is well established that gold binds MT with an affinity comparable to that of cadmium [183], although the specificity is much lower. These omissions indicate potential for optimisation of MT as a TEM label, as well as for fusion to GFP for use in CLEM.

1.6 Objectives of this Thesis

The use of cryoTEM to elucidate the superstructure of a *de novo* designed self-assembling peptide fibre (SAF) system is described in Chap. 4. The SAF system is the most well developed *de novo* designed fibre system currently in existence. With the elucidation of the formation pathway and numerous mutations to it, our understanding of the fibre has increased. The final step is to elucidate the ultrastructure of the fibre itself. Although composed of coiled coils, the method of lateral and longitudinal assembly has not been determined. By using cryoTEM to view the protein directly it is possible to elucidate the packing regime within the fibres, leading to a greater understanding of the rules governing coiled-coil fibre assembly and modulation.

The design of probes suitable for CLEM is described in Chap. 5. Two separate strategies are attempted with the primary aim of describing a genetically encoded protein that can be used as a clonable tag for use in CLEM. Firstly, the concatenation of MT and fusion to EGFP would allow visualisation of tagged proteins in both the LM and EM. Formation of protein-mediated electron-dense nanoparticles will be characterised and imaged *in vitro* prior to demonstration *in vivo*. The protein will be localised and imaged *in vivo* in both the LM and EM. Secondly, the use of a *de novo* binding protein, FKBP12(F36V), as a potential localisation motif for nanogold is described.

References

1. M. Uecker, S. Zhang, D. Voit, A. Karaus, K.-D. Merboldt, J. Frahm, Real-time MRI at a resolution of 20 ms. NMR Biomed. **23**(8), 986–994 (2010)
2. M. Fernandez-Suarez, A.Y. Ting, Fluorescent probes for super-resolution imaging in living cells. Nat. Rev. Mole. Cell Biol. **9**(12), 929–943 (2008)
3. L. Schermelleh, R. Heintzmann, H. Leonhardt, A guide to super-resolution fluorescence microscopy. J. Cell Biol. **190**(2), 165–175 (2010)
4. P.J. Keller, A.D. Schmidt, J. Wittbrodt, E.H.K. Stelzer, Reconstruction of zebrafish early embryonic development by scanned light sheet microscopy. Science **322**(5904), 1065–1069 (2008)
5. A. Egner, S.W. Hell, Fluorescence microscopy with super-resolved optical sections. Trends Cell Biol. **15**(4), 207–215 (2005)
6. J.A. Steyer, W. Almers, A real-time view of life within 100 nm of the plasma membrane. Nat. Rev. Mol. Cell Biol. **2**(4), 268–275 (2001)
7. S.W. Hell, Toward fluorescence nanoscopy. Nat. Biotechnol. **21**(11), 1347–1355 (2003)
8. G. Perinetti, T. Müller, A. Spaar, R. Polishchuk, A. Luini, A. Egner, Correlation of 4Pi and electron microscopy to study transport through single Golgi stacks in living cells with super resolution. Traffic **10**(4), 379–391 (2009)
9. E. Betzig, G.H. Patterson, R. Sougrat, O.W. Lindwasser, S. Olenych, J.S. Bonifacino, M.W. Davidson, J. Lippincott-Schwartz, H.F. Hess, Imaging intracellular fluorescent proteins at nanometer resolution. Science **313**(5793), 1642–1645 (2006)
10. M.J. Rust, M. Bates, X. Zhuang, Sub-diffraction-limit imaging by stochastic optical reconstruction microscopy (STORM). Nat. Meth. **3**(10), 793–795 (2006)
11. B. Huang, W. Wang, M. Bates, X. Zhuang, Three-dimensional super-resolution imaging by stochastic optical reconstruction microscopy. Science **319**(5864), 810–813 (2008)
12. G. Shtengel, J.A. Galbraith, C.G. Galbraith, J. Lippincott-Schwartz, J.M. Gillette, S. Manley, R. Sougrat, C.M. Waterman, P. Kanchanawong, M.W. Davidson, R.D. Fetter, H.F. Hess, Interferometric fluorescent super-resolution microscopy resolves 3D cellular ultrastructure. PNAS **106**(9), 3125–3130 (2009)
13. R. Ando, H. Hama, M. Yamamoto-Hino, H. Mizuno, A. Miyawaki, An optical marker based on the UV-induced green-to-red photoconversion of a fluorescent protein. PNAS **99**(20), 12651–12656 (2002)
14. P. Wilmann, K. Turcic, J. Battad, M. Wilce, R. Devenish, M. Prescott, J. Rossjohn, The 1.7Å crystal structure of Dronpa: A photoswitchable green fluorescent protein. J. Mol. Biol. **364**(2), 213–224 (2006)
15. K.I. Willig, R.R. Kellner, R. Medda, B. Hein, S. Jakobs, S.W. Hell, Nanoscale resolution in GFP-based microscopy. Nat. Meth. **3**(9), 721–723 (2006)
16. T. Grotjohann, I. Testa, M. Leutenegger, H. Bock, N.T. Urban, F. Lavoie-Cardinal, K.I. Willig, C. Eggeling, S. Jakobs, S.W. Hell, Diffraction-unlimited all-optical imaging and writing with a photochromic GFP. Nature, vol. advance online publication (2011)
17. V. Westphal, M.A. Lauterbach, A. Di Nicola, S.W. Hell, Dynamic far-field fluorescence nanoscopy. New J. Phys. **9**(12), 435+ (2007)
18. V. Westphal, S.O. Rizzoli, M.A. Lauterbach, D. Kamin, R. Jahn, S.W. Hell, Video-rate far-field optical nanoscopy dissects synaptic vesicle movement. Science **320**(5873), 246–249 (2008)
19. W. Min, S. Lu, S. Chong, R. Roy, G.R. Holtom, X.S. Xie, Imaging chromophores with undetectable fluorescence by stimulated emission microscopy. Nature **461**(7267), 1105–1109 (2009)
20. O. Shimomura, F.H. Johnson, Y. Saiga, Extraction, purification and properties of aequorin, a bioluminescent protein from the luminous hydromedusan, Aequorea. J. Cell. Comp. Physiol. **59**, 223–239 (1962)
21. H. Morise, O. Shimomura, F.H. Johnson, J. Winant, Intermolecular energy transfer in the bioluminescent system of Aequorea. Biochemistry **13**(12), 2656–2662 (1974)

22. M. Ormö, A.B. Cubitt, K. Kallio, L.A. Gross, R.Y. Tsien, S.J. Remington, Crystal structure of the Aequorea victoria green fluorescent protein. Science **273**(5280), 1392–1395 (1996)
23. F. Yang, L.G. Moss, G.N. Phillips, The molecular structure of green fluorescent protein. Nat. Biotechnol. **14**(10), 1246–1251 (1996)
24. S.E. Jackson, T.D. Craggs, J.-R. Huang, Understanding the folding of GFP using biophysical techniques. Expert Rev. Proteomics **3**(5), 545–559 (2006)
25. R. Heim, A.B. Cubitt, R.Y. Tsien, Improved green fluorescence. Nature **373**(6516), 663–664 (1995)
26. B. Cormack, FACS-optimized mutants of the green fluorescent protein (GFP). Gene **173**(1), 33–38 (1996)
27. T.T. Yang, L. Cheng, S.R. Kain, Optimized codon usage and chromophore mutations provide enhanced sensitivity with the green fluorescent protein. Nucleic Acids Res. **24**(22), 4592–4593 (1996)
28. R. Heim, D.C. Prasher, R.Y. Tsien, Wavelength mutations and posttranslational autoxidation of green fluorescent protein. PNAS **91**(26), 12501–12504 (1994)
29. M.V. Matz, A.F. Fradkov, Y.A. Labas, A.P. Savitsky, A.G. Zaraisky, M.L. Markelov, S.A. Lukyanov, Fluorescent proteins from nonbioluminescent Anthozoa species. Nat. Biotechnol. **17**(10), 969–973 (1999)
30. N.C. Shaner, R.E. Campbell, P.A. Steinbach, B.N.G. Giepmans, A.E. Palmer, R.Y. Tsien, Improved monomeric red, orange and yellow fluorescent proteins derived from Discosoma sp. red fluorescent protein. Nat. Biotechnol. **22**(12), 1567–1572 (2004)
31. G.H. Patterson, J. Lippincott-Schwartz, A photoactivatable GFP for selective photolabeling of proteins and cells. Science **297**(5588), 1873–1877 (2002)
32. H.P. Erickson, A. Klug, Measurement and compensation of defocusing and aberrations by Fourier processing of electron micrographs. Philos. Trans. Royal Soc. London Ser. B Biol. Sci. **261**(837), 105–118 (1971)
33. Tsuda Tanaka, Terauchi, Tsuno, Kaneyama, Honda, and Ishida, "A new 200 kV Ω-filter electron microscope". J. Microsc. **194**(1), 219–227 (1999)
34. D.B. Williams, C.B. Carter, *Transmission Electron Microscopy: A Textbook for Materials Science* (Spinger, New York, 2009)
35. M. Haider, A spherical-aberration-corrected 200kV transmission electron microscope. Ultramicroscopy **75**(1), 53–60 (1998)
36. P.E. Batson, N. Dellby, O.L. Krivanek, Sub-ångstrom resolution using aberration corrected electron optics. Nature **418**(6898), 617–620 (2002)
37. J.E. Evans, C. Hetherington, A. Kirkland, L.-Y. Y. Chang, H. Stahlberg, N. Browning, Low-dose aberration corrected cryo-electron microscopy of organic specimens. Ultramicroscopy **108**(12), 1636–1644 (2008)
38. J.L. Hutchison, J.M. Titchmarsh, D.J.H. Cockayne, R.C. Doole, C.J.D. Hetherington, A.I. Kirkland, H. Sawada, A versatile double aberration-corrected, energy filtered HREM/STEM for materials science. Ultramicroscopy **103**(1), 7–15 (2005)
39. K. Nagayama, R. Danev, Phase contrast electron microscopy: development of thin-film phase plates and biological applications. Philos. Trans. Roy. Soc. London Ser. B Biol. Sci. **363**(1500), 2153–2162 (2008)
40. R. Danev, Transmission electron microscopy with Zernike phase plate. Ultramicroscopy **88**(4), 243–252 (2001)
41. E. Majorovits, B. Barton, K. Schultheiß, F. Pérez-Willard, D. Gerthsen, R.R. Schröder, Optimizing phase contrast in transmission electron microscopy with an electrostatic (Boersch) phase plate. Ultramicroscopy **107**(2–3), 213–226 (2007)
42. R. Danev, K. Nagayama, Phase plates for transmission electron microscopy. Meth. Enzymol. **481**, 343–369 (2010)
43. J.V. Small, Organization of actin in the leading edge of cultured cells: influence of osmium tetroxide and dehydration on the ultrastructure of actin meshworks. J. Cell Biol. **91**(3), Pt 1, 695–705 (1981)

44. J.L. Murk, G. Posthuma, A.J. Koster, H.J. Geuze, A.J. Verkleij, M.J. Kleijmeer, B.M. Humbel, Influence of aldehyde fixation on the morphology of endosomes and lysosomes: quantitative analysis and electron tomography. J. Microsc. **212**(Pt 1), 81–90 (2003)
45. E. Brown, J. Mantell, D. Carter, G. Tilly, P. Verkade, Studying intracellular transport using high-pressure freezing and Correlative Light Electron Microscopy. Semin. Cell Dev. Biol. **20**(8), 910–919 (2009)
46. P. Verkade, Moving EM: the Rapid Transfer System as a new tool for correlative light and electron microscopy and high throughput for high-pressure freezing. J. Microsc. **230**(Pt 2), 317–328 (2008)
47. H. Moor, *Theory and Practice of High Pressure Freezing* (Springer, New York, 1987), pp. 175–191
48. R.A. Steinbrecht, M. Muller, *Freeze-Substitution and Freeze-Drying*, Chap. 7. (Springer, New York, 1987)
49. A. Al-Amoudi, J.-J. J. Chang, A. Leforestier, A. McDowall, L.M.M. Salamin, L.P. Norlén, K. Richter, N.S.S. Blanc, D. Studer, J. Dubochet, Cryo-electron microscopy of vitreous sections. EMBO J. **23**(18), 3583–3588 (2004)
50. A. Al-Amoudi, J. Dubochet, H. Gnaegi, W. Lüthi, D. Studer, An oscillating cryo-knife reduces cutting-induced deformation of vitreous ultrathin sections. J. Microsc. **212**(1), 26–33 (2003)
51. A. Rigort, F.J.B. Bäuerlein, A. Leis, M. Gruska, C. Hoffmann, T. Laugks, U. Böhm, M. Eibauer, H. Gnaegi, W. Baumeister, J.M. Plitzko, Micromachining tools and correlative approaches for cellular cryo-electron tomography. J. Struct. Biol. **172**(2), 169–179 (2010)
52. H.-W. Wang, J. Wang, F. Ding, K. Callahan, M.A. Bratkowski, J.S. Butler, E. Nogales, A. Ke, Architecture of the yeast Rrp44exosome complex suggests routes of RNA recruitment for 3 end processing. PNAS **104**(43), 16844–16849 (2007)
53. A.W. Robards, U.B. Sleytr, *Low Temperature Methods in Biological Electron Microscopy*, vol. 10 (Elsevier, New York, 1985)
54. J. Dubochet, A.W. McDowall, Vitrification of pure water for electron microscopy. J. Microsc. **124**, RP3–RP4 (1981)
55. N. Roos, A.J. Morgan, *Cryopreparation of Thin Biological Specimens for Electron Microscopy: Methods and Applications* (Oxford University Press, Oxford, 1990)
56. J.P. Buban, Q. Ramasse, B. Gipson, N.D. Browning, H. Stahlberg, High-resolution low-dose scanning transmission electron microscopy. J. Electron Microsc. **59**(2), 103–112 (2010)
57. L.A. Baker, J.L. Rubinstein, Radiation damage in electron cryomicroscopy. Methods Enzymol. **481**, 371–388 (2010)
58. P.A. Penczek, J. Zhu, R. Schroder, J. Frank, Three dimensional reconstruction with contrast transfer compensation from defocus series. Scan. Microsopy **11**, 147–154 (1997)
59. T.R. Shaikh, H. Gao, W.T. Baxter, F.J. Asturias, N. Boisset, A. Leith, J. Frank, SPIDER image processing for single-particle reconstruction of biological macromolecules from electron micrographs. Nat. Protoc. **3**(12), 1941–1974 (2008)
60. T. Kato, R.P. Goodman, C.M. Erben, A.J. Turberfield, K. Namba, High-resolution structural analysis of a DNA nanostructure by cryoEM. Nano Lett. **9**(7), 2747–2750 (2009)
61. K. Yonekura, C. Toyoshima, S. Maki-Yonekura, K. Namba, GUI programs for processing individual images in early stages of helical image reconstruction for high-resolution structure analysis. J. Struct. Biol. **144**(1–2), 184–194 (2003)
62. X. Zhang, E. Settembre, C. Xu, P.R. Dormitzer, R. Bellamy, S.C. Harrison, N. Grigorieff, Near-atomic resolution using electron cryomicroscopy and single-particle reconstruction. PNAS **105**(6), 1867–1872 (2008)
63. R. Henderson, J.M. Baldwin, T.A. Ceska, F. Zemlin, E. Beckmann, K.H. Downing, Model for the structure of bacteriorhodopsin based on high-resolution electron cryo-microscopy. J. Mol. Biol. **213**(4), 899–929 (1990)
64. E. Egelman, The iterative helical real space reconstruction method: Surmounting the problems posed by real polymers. J. Struct. Biol. **157**(1), 83–94 (2007)
65. J. Frauenfeld, J. Gumbart, E.O. Sluis, S. Funes, M. Gartmann, B. Beatrix, T. Mielke, O. Berninghausen, T. Becker, K. Schulten, R. Beckmann, Cryo-EM structure of the ribosome-SecYE complex in the membrane environment. Nat. Struct. Mol. Biol. **18**(5), 614–621 (2011)

66. J.O. Ortiz, F. Förster, J. Kürner, A.A. Linaroudis, W. Baumeister, Mapping 70S ribosomes in intact cells by cryoelectron tomography and pattern recognition. J. Struct. Biol. **156**(2), 334–341 (2006)
67. N.R. Voss, C.S. Potter, R. Smith, B. Carragher, Software tools for molecular microscopy: an open-text Wikibook. Methods Enzymol. **482**, 381–392 (2010)
68. D.J. De Rosier, A. Klug, Reconstruction of three dimensional structures from electron micrographs. Nature **217**(5124), 130–134 (1968)
69. T. Dang, S. Farah, A. Gast, C. Robertson, B. Carragher, E. Egelman, E. Wilsonkubalek, Helical crystallization on lipid nanotubes: Streptavidin as a model protein. J. Struct. Biol. **150**(1), 90–99 (2005)
70. R. Diaz, W.J. Rice, D.L. Stokes, Fourier-Bessel reconstruction of helical assemblies. Methods Enzymol. **482**, 131–165 (2010)
71. J. Harford, J. Squire, Time-resolved diffraction studies of muscle using synchrotron radiation. Rep. Prog. Phys. **60**, 1723–1787 (1997)
72. M.F. Moody, *Image Analysis of Electron Micrographs* (Academic Press, New York, 1990), pp. 145–287
73. M. Hu, L. Qian, R.P. Briñas, E.S. Lymar, L. Kuznetsova, J.F. Hainfeld, Gold nanoparticle-protein arrays improve resolution for cryo-electron microscopy. J. Struct. Biol. **161**(1), 83–91 (2008)
74. D.G. Morgan, Q.M. Ramasse, N.D. Browning, Application of two-dimensional crystallography and image processing to atomic resolution Z-contrast images. J. Electron Microsc. **58**(3), 223–244 (2009)
75. R. Henderson, J. Baldwin, K. Downing, J. Lepault, F. Zemlin, Structure of purple membrane from halobacterium halobium: recording, measurement and evaluation of electron micrographs at 3.5 Å resolution. Ultramicroscopy **19**(2), 147–178 (1986)
76. B. Gipson, X. Zeng, H. Stahlberg, 2dx_merge: Data management and merging for 2D crystal images. J. Struct. Biol. **160**(3), 375–384 (2007)
77. A.J. Koster, R. Grimm, D. Typke, R. Hegerl, A. Stoschek, J. Walz, W. Baumeister, Perspectives of molecular and cellular electron tomography. J. Struct. Biol. **120**(3), 276–308 (1997)
78. D. Hanein, N. Volkmann, *Correlative Light & Electron Microscopy*, vol. 82, (Elsevier, 2011), pp. 91–99
79. L. R. Otterbein, P. Graceffa, R. Dominguez, The crystal structure of uncomplexed actin in the ADP state. Science **293**(5530), 708–711 (2001)
80. M.A. Rould, Q. Wan, P.B. Joel, S. Lowey, K.M. Trybus, Crystal structures of expressed non-polymerizable monomeric actin in the ADP and ATP states. J. Biol. Chem. **281**(42), 31909–31919 (2006)
81. K. Murakami, T. Yasunaga, T.Q.P. Noguchi, Y. Gomibuchi, K.X. Ngo, T.Q.P. Uyeda, T. Wakabayashi, Structural basis for actin assembly, activation of ATP hydrolysis, and delayed phosphate release. Cell **143**(2), 275–287 (2010)
82. T. Fujii, A.H. Iwane, T. Yanagida, K. Namba, Direct visualization of secondary structures of F-actin by electron cryomicroscopy. Nature **467**(7316), 724–728 (2010)
83. R. Dominguez, K.C. Holmes, Actin structure and function. Ann. Rev. Biophys. **40**(1), 169–186 (2011)
84. H. Li, D. Derosier, W. Nicholson, E. Nogales, K. Downing, Microtubule Structure at 8 Å Resolution. Structure **10**(10), 1317–1328 (2002)
85. E. Nogales, S.G. Wolf, K.H. Downing, Structure of the alpha beta tubulin dimer by electron crystallography. Nature **391**(6663), 199–203 (1998)
86. J. Löwe, H. Li, K.H. Downing, E. Nogales, Refined structure of alpha beta-tubulin at 3.5 A resolution. J. Mol. Biol. **313**(5), 1045–1057 (2001)
87. H. Sui, K.H. Downing, Structural basis of interprotofilament interaction and lateral deformation of microtubules. Structure **18**(8), 1022–1031 (2010)
88. H. Herrmann, H. Bar, L. Kreplak, S.V. Strelkov, U. Aebi, Intermediate filaments: from cell architecture to nanomechanics. Nat. Rev. Mol. Cell Biol. **8**(7), 562–573 (2007)

89. R. Kirmse, S. Portet, N. Mücke, U. Aebi, H. Herrmann, J. Langowski, A quantitative kinetic model for the in vitro assembly of intermediate filaments from tetrameric vimentin. J. Biol. Chem. **282**(25), 18563–18572 (2007)
90. H. Bär, N. Mücke, A. Kostareva, G. Sjöberg, U. Aebi, H. Herrmann, Severe muscle disease-causing desmin mutations interfere with in vitro filament assembly at distinct stages. PNAS **102**(42), 15099–15104 (2005)
91. S.V. Strelkov, H. Herrmann, U. Aebi, Molecular architecture of intermediate filaments. BioEssays **25**(3), 243–251 (2003)
92. K.N. Goldie, T. Wedig, A.K. Mitra, U. Aebi, H. Herrmann, A. Hoenger, Dissecting the 3-D structure of vimentin intermediate filaments by cryo-electron tomography. J. Struct. Biol. **158**(3), 378–385 (2007)
93. L. Norlén, S. Masich, K.N. Goldie, A. Hoenger, Structural analysis of vimentin and keratin intermediate filaments by cryo-electron tomography. Exp. Cell Res. **313**(10), 2217–2227 (2007)
94. S. Nicolet, H. Herrmann, U. Aebi, S.V. Strelkov, Atomic structure of vimentin coil 2. J. Struct. Biol. **170**(2), 369–376 (2010)
95. H. Herrmann, U. Aebi, Intermediate filaments: molecular structure, assembly mechanism, and integration into functionally distinct intracellular scaffolds. Ann. Rev. Biochem. **73**(1), 749–789 (2004)
96. N.R. Watts, L.N. Jones, N. Cheng, J.S. Wall, D.A. Parry, A.C. Steven, Cryo-electron microscopy of trichocyte (hard alpha-keratin) intermediate filaments reveals a low-density core. J. Struct. Biol. **137**(1–2), 109–118 (2002)
97. M.T. Cabeen, C.J. Wagner, The Bacterial Cytoskeleton. Ann. Rev. Genetics **44**(1), 365–392 (2010)
98. T. Minamino, K. Imada, K. Namba, Molecular motors of the bacterial flagella. Current Opin. Struct. Biol. **18**(6), 693–701 (2008)
99. Y. Mimori, I. Yamashita, K. Murata, Y. Fujiyoshi, K. Yonekura, C. Toyoshima, K. Namba, The structure of the R-type straight flagellar filament of salmonella at 9 Å resolution by electron cryomicroscopy. J. Mol. Biol. **249**(1), 69–87 (1995)
100. K. Yonekura, S. Maki-Yonekura, K. Namba, Complete atomic model of the bacterial flagellar filament by electron cryomicroscopy. Nature **424**(6949), 643–650 (2003)
101. K. Yonekura, S. Maki-Yonekura, K. Namba, Building the atomic model for the bacterial flagellar filament by electron cryomicroscopy and image analysis. Structure **13**(3), 407–412 (2005)
102. F.A. Samatey, K. Imada, S. Nagashima, F. Vonderviszt, T. Kumasaka, M. Yamamoto, K. Namba, Structure of the bacterial flagellar protofilament and implications for a switch for supercoiling. Nature **410**(6826), 331–337 (2001)
103. S. Maki-Yonekura, K. Yonekura, K. Namba, Conformational change of flagellin for polymorphic supercoiling of the flagellar filament. Nat. Struct. Mol. Biol. **17**(4), 417–422 (2010)
104. A.S. Cohen, E. Calkins, Electron microscopic observations on a fibrous component in amyloid of diverse origins. Nature **183**(4669), 1202–1203 (1959)
105. M. Sunde, L.C. Serpell, M. Bartlam, P.E. Fraser, M.B. Pepys, C.C. Blake, Common core structure of amyloid fibrils by synchrotron X-ray diffraction. J. Mol. Biol. **273**(3), 729–739 (1997)
106. J.L. Jiménez, Guijarro, E. Orlova, J. Zurdo, C.M. Dobson, M. Sunde, H.R. Saibil, Cryo-electron microscopy structure of an SH3 amyloid fibril and model of the molecular packing, EMBO J. **18**(4), 815–821 (1999)
107. J.L. Jiménez, G. Tennent, M. Pepys, H.R. Saibil, Structural diversity of ex vivo amyloid fibrils studied by cryo-electron microscopy. J. Mol. Biol. **311**(2), 241–247 (2001)
108. J.L. Jiménez, E.J. Nettleton, M. Bouchard, C.V. Robinson, C.M. Dobson, H.R. Saibil, The protofilament structure of insulin amyloid fibrils. PNAS **99**(14), 9196–9201 (2002)
109. T.P.J. Knowles, M.J. Buehler, Nanomechanics of functional and pathological amyloid materials. Nat. Nanotechnol. **6**(8), 469–479 (2011)

110. R. Zhang, X. Hu, H. Khant, S.J. Ludtke, W. Chiu, M.F. Schmid, C. Frieden, J.-M. M. Lee, Interprotofilament interactions between Alzheimer's Abeta1-42 peptides in amyloid fibrils revealed by cryoEM. PNAS **106**(12), 4653–4658 (2009)
111. M. Schmidt, C. Sachse, W. Richter, C. Xu, M. Fändrich, N. Grigorieff, Comparison of Alzheimer Abeta(1–40) and Abeta(1–42) amyloid fibrils reveals similar protofilament structures. PNAS **106**(47), 19813–19818 (2009)
112. M.R. Sawaya, S. Sambashivan, R. Nelson, M.I. Ivanova, S.A. Sievers, M.I. Apostol, M.J. Thompson, M. Balbirnie, J.J. Wiltzius, H.T. McFarlane, A. Ø. Madsen, C. Riekel, D. Eisenberg, Atomic structures of amyloid cross-beta spines reveal varied steric zippers. Nature **447**(7143), 453–457 (2007)
113. J.J.W. Wiltzius, S.A. Sievers, M.R. Sawaya, D. Cascio, D. Popov, C. Riekel, D. Eisenberg, Atomic structure of the cross-beta spine of islet amyloid polypeptide (amylin). Protein Sci. **17**(9), 1467–1474 (2008)
114. H.E. White, J.L. Hodgkinson, T.R. Jahn, S. Cohen-Krausz, W.S. Gosal, S. Müller, E.V. Orlova, S.E. Radford, H.R. Saibil, Globular tetramers of β_2-microglobulin assemble into elaborate amyloid fibrils. J. Mol. Biol. **389**(1), 48–57 (2009)
115. M.H. Tattum, S. Cohen-Krausz, A. Khalili-Shirazi, G.S. Jackson, E.V. Orlova, J. Collinge, A.R. Clarke, H.R. Saibil, Elongated oligomers assemble into mammalian PrP amyloid fibrils. J. Mol. Biol. **357**(3), 975–985 (2006)
116. C. Liu, M.R. Sawaya, D. Eisenberg, β_2-microglobulin forms three-dimensional domain-swapped amyloid fibrils with disulfide linkages. Nat. Struct. Mol. Biol. **18**(1), 49–55 (2011)
117. A.L. Boyle, D.N. Woolfson, De novo designed peptides for biological applications. Chem. Soc. Rev. **40**(8), 4295–4306 (2011)
118. D.N. Woolfson, Z.N. Mahmoud, More than just bare scaffolds: towards multi-component and decorated fibrous biomaterials. Chem. Soc. Rev. **39**(9), 3464–3479 (2010)
119. R.V. Ulijn, A.M. Smith, Designing peptide based nanomaterials. Chem. Soc. Rev. **37**(4), 664–675 (2008)
120. M. Zhou, A.M. Smith, A.K. Das, N.W. Hodson, R.F. Collins, R.V. Ulijn, J.E. Gough, Self-assembled peptide-based hydrogels as scaffolds for anchorage-dependent cells. Biomaterials **30**(13), 2523–2530 (2009)
121. R.P. Nagarkar, R.A. Hule, D.J. Pochan, J.P. Schneider, De novo design of strand-swapped beta-hairpin hydrogels. J. Am. Chem. Soc. **130**(13), 4466–4474 (2008)
122. J.D. Hartgerink, E. Beniash, S.I. Stupp, Self-assembly and mineralization of peptide-amphiphile nanofibers. Science **294**(5547), 1684–1688 (2001)
123. M. Bruning, L. Kreplak, S. Leopoldseder, S.A. Muller, P. Ringler, L. Duchesne, D.G. Fernig, A. Engel, Z. Ucurum-Fotiadis, O. Mayans, Bipartite design of a self-fibrillating protein copolymer with nanopatterned peptide display capabilities. Nano Lett. **10**(11), 4533–4537 (2010)
124. J.C. Sinclair, K.M. Davies, C. Venien-Bryan, M.E.M. Noble, Generation of protein lattices by fusing proteins with matching rotational symmetry. Nat. Nanotechnol. vol. advance online, publication (2011)
125. R. Orbach, L. Adler-Abramovich, S. Zigerson, I. Mironi-Harpaz, D. Seliktar, E. Gazit, Self-assembled Fmoc-peptides as a platform for the formation of nanostructures and hydrogels. Biomacromolecules **10**(9), 2646–2651 (2009)
126. K. Rajagopal, M.S. Lamm, L.A. Haines-Butterick, D.J. Pochan, J.P. Schneider, Tuning the pH Responsiveness of beta-Hairpin Peptide Folding, Self-Assembly, and Hydrogel Material Formation. Biomacromolecules **10**(9), 2619–2625 (2009)
127. D.N. Woolfson, Building fibrous biomaterials from alpha-helical and collagen-like coiled-coil peptides. Biopolymers **94**(1), 118–127 (2010)
128. D.N. Woolfson, The design of coiled-coil structures and assemblies. Adv. Protein Chem. **70**, 79–112 (2005)
129. P.B. Harbury, T. Zhang, P.S. Kim, T. Alber, A switch between two-, three-, and four-stranded coiled coils in GCN4 leucine zipper mutants. Science **262**(5138), 1401–1407 (1993)

130. M.J. Pandya, G.M. Spooner, M. Sunde, J.R. Thorpe, A. Rodger, D.N. Woolfson, Sticky-end assembly of a designed peptide fiber provides insight into protein fibrillogenesis. Biochemistry **39**(30), 8728–8734 (2000)
131. S.C. Holmstrom, P.J.S. King, M.G. Ryadnov, M.F. Butler, S. Mann, D.N. Woolfson, Templating silica nanostructures on rationally designed self-assembled peptide fibers. Langmuir **24**(20), 11778–11783 (2008)
132. D. Papapostolou, A.M. Smith, E.D.T. Atkins, S.J. Oliver, M.G. Ryadnov, L.C. Serpell, D.N. Woolfson, Engineering nanoscale order into a designed protein fiber. PNAS **104**(26), 10853–10858 (2007)
133. A.M. Smith, E.F. Banwell, W.R. Edwards, M.J. Pandya, D.N. Woolfson, Engineering increased stability into self-assembled protein fibers. Adv. Funct. Mater. **16**(8), 1022–1030 (2006)
134. E.H. Bromley, K.J. Channon, P.J. King, Z.N. Mahmoud, E.F. Banwell, M.F. Butler, M.P. Crump, T.R. Dafforn, M.R. Hicks, J.D. Hirst, A. Rodger, D.N. Woolfson, Assembly pathway of a designed alpha-helical protein fiber. Biophys. J. **98**(8), 1668–1676 (2010)
135. G. Griffiths, B. Burke, J. Lucocq, *Fine Structure Immunocytochemistry* (Springer, New York, 1993)
136. G. Patterson, M. Davidson, S. Manley, J.L. Schwartz, Superresolution imaging using single-molecule localization. Ann. Rev. Phys. Chem. **61**(1), 345–367 (2010)
137. B. Hein, K.I. Willig, C.A. Wurm, V. Westphal, S. Jakobs, S.W. Hell, Stimulated emission depletion nanoscopy of living cells using SNAP-Tag fusion proteins. Biophys. J. **98**(1), 158–163 (2010)
138. A.V. Agronskaia, J.A. Valentijn, L.F. van Driel, C.T. Schneijdenberg, B.M. Humbel, P.M. van Bergen en Henegouwen, A.J. Verkleij, A.J. Koster, H.C. Gerritsen, Integrated fluorescence and transmission electron microscopy. J. Struct. Biol. **164**(2), 183–189 (2008)
139. L.F. van Driel, J.A. Valentijn, K.M. Valentijn, R.I. Koning, A.J. Koster, Tools for correlative cryo-fluorescence microscopy and cryo-electron tomography applied to whole mitochondria in human endothelial cells. Eur. J. Cell Biol. **88**(11), 669–684 (2009)
140. A. Sartori, R. Gatz, F. Beck, A. Rigort, W. Baumeister, J.M. Plitzko, Correlative microscopy: Bridging the gap between fluorescence light microscopy and cryo-electron tomography. J. Struct. Biol. **160**(2), 135–145 (2007)
141. C.L. Schwartz, V.I. Sarbash, F.I. Ataullakhanov, J.R. McIntosh, D. Nicastro, Cryo-fluorescence microscopy facilitates correlations between light and cryo-electron microscopy and reduces the rate of photobleaching. J. Microsc. **227**(Pt 2), 98–109 (2007)
142. L. van Driel, *Correlative Light and Electron Microscopy: Strategies and Applications* Ph.D. thesis, (Leiden University, 2011)
143. S. Lepper, M. Merkel, A. Sartori, M. Cyrklaff, F. Frischknecht, Rapid quantification of the effects of blotting for correlation of light and cryo-light microscopy images. J. Microsc. **238**(1), 21–26 (2010)
144. H. Schwarz, B.M. Humbel, Correlative light and electron microscopy using immunolabeled resin sections. in *Electron Microscopy : Methods and Protocols of Methods in Molecular Biology*, vol. 369, ed. by J. Kuo, Chap. 12, (Humana Press, Totowa, 2007), pp. 229–256
145. K.T. Tokuyasu, A technique for ultracryotomy of cell suspensions and tissues. J. Cell Biol. **57**(2), 551–565 (1973)
146. R.D. Powell, C.M.R. Halsey, J.F. Hainfeld, Combined fluorescent and gold immunoprobes: Reagents and methods for correlative light and electron microscopy. Microsc. Res. Technique **42**(1), 2–12 (1998)
147. J.R. van Weering, E. Brown, T.H. Sharp, J. Mantell, P.J. Cullen, P. Verkade, Intracellular membrane traffic at high resolution. Methods Cell Biol. **96**, 619–648 (2010)
148. E. van Donselaar, G. Posthuma, D. Zeuschner, B.M. Humbel, J.W. Slot, Immunogold labeling of cryosections from high-pressure frozen cells. Traffic (Copenhagen, Denmark) **8**(5), 471–485 (2007)
149. V. Oorschot, H. de Wit, W.G. Annaert, J. Klumperman, A novel flat-embedding method to prepare ultrathin cryosections from cultured cells in their in situ orientation. J. Histochemistry Cytochem. **50**(8), 1067–1080 (2002)

150. C. van Rijnsoever, V. Oorschot, J. Klumperman, Correlative light-electron microscopy (CLEM) combining live-cell imaging and immunolabeling of ultrathin cryosections. Nat. Methods **5**(11), 973–980 (2008)
151. J.F. Hainfeld, F.R. Furuya, A 1.4-nm gold cluster covalently attached to antibodies improves immunolabeling. J. Histochemistry Cytochem. **40**(2), 177–184 (1992)
152. J.M. Robinson, D.D. Vandré, Efficient immunocytochemical labeling of leukocyte microtubules with FluoroNanogold: An important tool for correlative microscopy. J. Histochemistry Cytochem. **45**(5), 631–642 (1997)
153. B.N.G. Giepmans, T.J. Deerinck, B.L. Smarr, Y.Z. Jones, M.H. Ellisman, Correlated light and electron microscopic imaging of multiple endogenous proteins using Quantum dots. Nat. Methods **2**(10), 743–749 (2005)
154. X. Michalet, F. Pinaud, T.D. Lacoste, M. Dahan, M.P. Bruchez, P.A. Alivisatos, S. Weiss, Properties of fluorescent semiconductor nanocrystals and their application to biological labeling. Single Mol. **2**(4), 261–276 (2001)
155. B.N.G. Giepmans, S.R. Adams, M.H. Ellisman, R.Y. Tsien, The fluorescent toolbox for assessing protein location and function. Science **312**(5771), 217–224 (2006)
156. J. Jaiswal, S. Simon, Potentials and pitfalls of fluorescent quantum dots for biological imaging. Trends Cell Biol. **14**(9), 497–504 (2004)
157. R. Nisman, G. Dellaire, Y. Ren, R. Li, D.P. Bazett-Jones, Application of quantum dots as probes for correlative fluorescence, conventional, and energy-filtered transmission electron microscopy. J. Histochemistry Cytochem. **52**(1), 13–18 (2004)
158. B.A. Griffin, S.R. Adams, R.Y. Tsien, Specific covalent labeling of recombinant protein molecules inside live cells. Science **281**(5374), 269–272 (1998)
159. S.R. Adams, R.E. Campbell, L.A. Gross, B.R. Martin, G.K. Walkup, Y. Yao, J. Llopis, R.Y. Tsien, New biarsenical ligands and tetracysteine motifs for protein labeling in vitro and in vivo: Synthesis and biological applications. J. Am. Chem. Soc. **124**(21), 6063–6076 (2002)
160. B.A. Griffin, S.R. Adams, J. Jones, R.Y. Tsien, Fluorescent labeling of recombinant proteins in living cells with FlAsH. Methods Enzymol. **327**, 565–578 (2000)
161. G. Gaietta, T.J. Deerinck, S.R. Adams, J. Bouwer, O. Tour, D.W. Laird, G.E. Sosinsky, R.Y. Tsien, M.H. Ellisman, Multicolor and electron microscopic imaging of connexin trafficking. Science **296**(5567), 503–507 (2002)
162. J. Lanman, J. Crum, T.J. Deerinck, G.M. Gaietta, A. Schneemann, G.E. Sosinsky, M.H. Ellisman, J.E. Johnson, Visualizing flock house virus infection in Drosophila cells with correlated fluorescence and electron microscopy. J. Struct. Biol. **161**(3), 439–446 (2008)
163. G.M. Gaietta, B.N. Giepmans, T.J. Deerinck, W.B. Smith, L. Ngan, J. Llopis, S.R. Adams, R.Y. Tsien, M.H. Ellisman, Golgi twins in late mitosis revealed by genetically encoded tags for live cell imaging and correlated electron microscopy. PNAS **103**(47), 17777–17782 (2006)
164. E.Z. Monosov, T.J. Wenzel, G.H. Luers, J.A. Heyman, S. Subramani, Labeling of peroxisomes with green fluorescent protein in living P. pastoris cells. J. Histochemistry Cytochem. **44**(6), 581–589 (1996)
165. M. Grabenbauer, W.J.C. Geerts, J. Fernadez-Rodriguez, A. Hoenger, A.J. Koster, T. Nilsson, Correlative microscopy and electron tomography of GFP through photooxidation. Nat. Methods **2**(11), 857–862 (2005)
166. C. Meißlitzer-Ruppitsch, M. Vetterlein, H. Stangl, S. Maier, J. Neumüller, M. Freissmuth, M. Pavelka, A. Ellinger, Electron microscopic visualization of fluorescent signals in cellular compartments and organelles by means of DAB-photoconversion. Histochemistry Cell Biol. **130**(2), 407–419 (2008)
167. X. Shu, A. Royant, M.Z. Lin, T.A. Aguilera, V. Lev-Ram, P.A. Steinbach, R.Y. Tsien, Mammalian expression of infrared fluorescent proteins engineered from a bacterial phytochrome. Science **324**(5928), 804–807 (2009)
168. X. Shu, V. Lev-Ram, T.J. Deerinck, Y. Qi, E.B. Ramko, M.W. Davidson, Y. Jin, M.H. Ellisman, R.Y. Tsien, A genetically encoded tag for correlated light and electron microscopy of intact cells, tissues, and organisms. PLoS Biol. **9**(4), e1001041+ (2011)

169. Q. Wang, C.P. Mercogliano, J. Löwe, A ferritin-based label for cellular electron cryotomography. Structure **19**(2), 147–154 (2011)
170. T. Clackson, W. Yang, L.W. Rozamus, M. Hatada, J.F. Amara, C.T. Rollins, L.F. Stevenson, S.R. Magari, S.A. Wood, N.L. Courage, X. Lu, F. Cerasoli, M. Gilman, D.A. Holt, Redesigning an FKBP-ligand interface to generate chemical dimerizers with novel specificity. PNAS **95**(18), 10437–10442 (1998)
171. J.H. Kaegi, A. Schaeffer, Biochemistry of metallothionein. Biochemistry **27**(23), 8509–8515 (1988)
172. A.T. Miles, G.M. Hawksworth, J.H. Beattie, V. Rodilla, Induction, regulation, degradation, and biological significance of mammalian metallothioneins. Crit. Rev. Biochem. Mol. Biol. **35**(1), 35–70 (2000)
173. N. Romero-Isart, L.T. Jensen, O. Zerbe, D.R. Winge, M. Vasak, Engineering of metallothionein-3 neuroinhibitory activity into the inactive isoform metallothionein-1. J. Biol. Chem. **277**(40), 37023–37028 (2002)
174. K.B. Nielson, C.L. Atkin, D.R. Winge, Distinct metal-binding configurations in metallothionein. J. Biol. Chem. **260**(9), 5342–5350 (1985)
175. C.P. Mercogliano, D.J. Derosier, Gold nanocluster formation using metallothionein: Mass spectrometry and electron microscopy. J. Mol. Biol. **355**(2), 211–223 (2006)
176. C.P. Mercogliano, D.J. DeRosier, Concatenated metallothionein as a clonable gold label for electron microscopy. J. Struct. Biol. **160**(1), 70–82 (2007)
177. Y. Nishino, T. Yasunaga, A. Miyazawa, A genetically encoded metallothionein tag enabling efficient protein detection by electron microscopy. J. Electron Microsc. **56**(3), 93–101 (2007)
178. Y. Fukunaga, A. Hirase, H. Kim, N. Wada, Y. Nishino, A. Miyazawa, Electron microscopic analysis of a fusion protein of postsynaptic density-95 and metallothionein in cultured hippocampal neurons. J. Electron Microsc. **56**(4), 119–129 (2007)
179. E. Diestra, J. Fontana, P. Guichard, S. Marco, C. Risco, Visualization of proteins in intact cells with a clonable tag for electron microscopy. J. Struct. Biol. **165**(3), 157–168 (2009)
180. E. Diestra, B. Cayrol, V. Arluison, C. Risco, Cellular electron microscopy imaging reveals the localization of the Hfq protein close to the bacterial membrane. PLoS ONE **4**(12), e8301+ (2009)
181. C. Gammons, Y. Yu, A. Williamsjones, The disproportionation of gold(I) chloride complexes at 25 to 200°C. Geochim. Cosmochim. Acta **61**(10), 1971–1983 (1997)
182. X. Lu, H.Y. Tuan, B.A. Korgel, Y. Xia, Facile synthesis of gold nanoparticles with narrow size distribution by using AuCl or AuBr as the precursor. Chemistry **14**(5), 1584–1591 (2008)
183. G. Schmitz, D.T. Minkel, D. Gingrich, C.F. Shaw, The binding of Gold(I) to metallothionein. J. Inorganic Biochem. **12**(4), 293–306 (1980)

Part II
Materials and Methods

Chapter 2
Experimental Methods

Unless otherwise stated, chemical reagents were purchased from Sigma-Aldrich and molecular biology reagents and enzymes were purchased from New England Biolabs (NEB). Solution recipes are found in Appendix B.

2.1 Cell Culture

2.1.1 Prokaryotic Cells

Bacterial cell cultures were grown in Luria Bertani (LB) broth and shaken at 37 °C at 250 rpm. *Escherichia coli* (*E. coli*) DH5α cells (Invitrogen) were used to maintain vectors, BL21(DE3) (Stratagene) cells were used as expression hosts. Glycerol stocks were prepared by resuspending the cell pellet in LB broth with 20 % glycerol. Stocks were stored at −80 °C.

2.1.1.1 Transforming Chemically Competent *Escherichia coli* Cells

DH5α and BL21 cells were transformed with 1–10 ng of DNA by heat-shock. Two different volumes of transformed cells were spread onto selective agar plates. Non-transformed cells were also spread as a negative control and pUC19-transformed cells were grown on ampicillin-agar plates as a positive control. Positive and negative controls confirmed the transformation competency and antibiotic stock viability, respectively.

T. H. Sharp, *Biomolecular Imaging at High Spatial and Temporal Resolution In Vitro and In Vivo*, Springer Theses, DOI: 10.1007/978-3-319-02159-1_2,

2.1.2 *Eukaryotic Cells*

HeLa cells were maintained in Dulbecco's modified Eagle's medium (DMEM) supplemented with 10 % fetal calf serum and 1 % Penicillin-Streptomycin. Cells were grown at 37 °C in 5 % CO_2. Cells were passaged when approximately 80 % confluency was reached. Old media was removed and cells were washed twice with phosphate-buffered saline solution (PBS). Detachment from the growing surface was achieved by incubation for 5 min at 37 °C in 1 mL Trypsin-EDTA prior to gentle agitation to further dislodge cells. Cells were resuspended in 5 mL DMEM and 1 mL of this solution was added to 20 mL of fresh DMEM and incubated at 37 °C.

2.1.2.1 Transfection of HeLa Cells

HeLa cells were transfected using either FuGENE (Roche) or Lipofectamine 2000 (Invitrogen) according to manufacturer's instructions. Both were used in combination with OptiMEM reduced-serum media (Invitrogen). After transfection, cells were incubated at 37 °C overnight.

2.1.3 *Apoptosis Assays*

Either MTT (Novagen) or AlamarBlue (Invitrogen) were used as per the manufacturers instructions.

2.2 Molecular Biology

2.2.1 *DNA Quantitation*

DNA concentration was calculated by measuring UV absorbance from 230–320 nm. Reference samples were 5 mM Tris-HCl pH 8. The concentration was calculated using the formula:

$$[\text{dsDNA}] = OD_{260} \times 50 \times \text{dilution factor} \tag{2.1}$$

DNA quality was assessed by comparing the A_{260} and A_{280} values:

$$\text{Quality}_{DNA} = \frac{A_{260}}{A_{280}} \tag{2.2}$$

A value of 1.5–2 was judged as pure.

2.2.2 Plasmid Purification

Plasmid DNA was purified used QIAGEN Plasmid Purification kits (Mini or Maxi). Volumes of buffer used were as recommended by the manufacturer.

2.2.3 Agarose Gels

Agarose gels were made by dissolving agarose powder in TBE buffer to the required concentration (between 0.8–1.5 %). Gels were run in TBE buffer at 100 V and visualised over a UV lamp after staining with Ethidium Bromide or GelRed™ (Biotium). The QIAGEN QIAquick Gel Extraction kit was used to purify DNA from the gel following the manufacturer's instructions.

2.2.4 Restriction Digests and Vector Ligations

Digestion by restriction enzymes was performed as recommended by NEB, adding 0.1 mg/mL Bovine Serum Albumin (BSA) as standard. Double digests were designed to function in the same buffer unless otherwise stated. Digestion time was 1–1.5 h, followed by denaturation at either 65 or 80 °C for 20 min and purification using QIAGEN QIAquick PCR purification columns or gel extraction.

The following formulae were used to calculate the amount of vector and insert required for ligation.
Molecular weight of DNA:

$$Mr_{DNA} = \mathrm{N^{o}{\cdot}bp} \times 660 \tag{2.3}$$

Concentration of DNA in fmoles/μL:

$$C_{DNA}(b) = \frac{\mathrm{mass}_{DNA}(\mathrm{ng}/\mu\mathrm{L})}{Mr_{DNA}} \times 10^{6} \tag{2.4}$$

For ligation, 30 fmoles each of vector and insert were added as standard. Incubation with 1 unit T4 DNA Ligase varied from 3 h at room temperature (21 °C) or 4 °C overnight for sticky-ended ligation, to 16 °C overnight for blunt-ended ligations. Ligase was not deactivated before transformation.

2.2.5 PCR, Gradient PCR and Colony PCR

Primers were designed by hand, verified using FastPCR (Primer Digital Ltd.) and synthesised by Eurofins MWG Operon (Ebersberg, Germany). Taq DNA polymerase was used for analytical PCR, whereas Vent$_R$ DNA polymerase was used when the product was to be sequenced or ligated. PCR generally followed the thermal cycle: initial melting 94 °C for 5 min, then; 30 × 94 °C for 1 min, Tm −5 °C for 45 s, 72 °C for 1 min, followed by final extension at 72 °C for 5 min. Gradient PCR used a similar cycle with annealing temperature varied from $T_m \pm 10$ °C, and the magnesium concentration varied from 2–6 mM. Colony PCR was performed to identify colonies with the correct ligation product. Selected colonies were picked with a sterile tip, swirled in deionised H_2O and deposited in selective LB Broth for culture. Samples were boiled at 99 °C for 5 min, centrifuged and 10 µL of the supernatant used for PCR. Purification of PCR samples was performed with PCR column purification kits (QIAGEN).

2.3 Gene Synthesis and Subcloning

Primer sequences are provided in Appendix B in the manner and order as they appear in the text.

2.3.1 Bacterial Expression Plasmids

To confirm successful subcloning of genes, each vector was sequenced by Eurofins MWG Operon (Ebersberg, Germany) using the Sanger dideoxy sequencing method. To simplify purification a His_6 tag was added to fusion proteins.

2.3.1.1 His_6-Tagged EGFP

A His_6 tag was added to fusion proteins to enable one-step purification. pET-45b(+) (Novagen) encodes an N-terminal His_6 sequence followed by a multiple cloning site (MCS). EGFP was subcloned into pET-45b(+) from pEGFP-C1 (Clontech) using the forward primer *His EGFP FWD*, which adds an MfeI site (underlined, Appendix B) before the EGFP start codon, and the reverse primer *His EGFP REV*, which retains the XhoI site (underlined) present in the MCS. Both insert and vector were digested with MfeI and XhoI and ligated as described to create pET-45b(+)-EGFP (Fig. 2.1).

2.3.1.2 His_6-Tagged EGFP-MT2

To create an inducible EGFP-MT2 fusion protein, two MT genes were subcloned adjacent to one another within the original pET-3d vector before subcloning the whole MT2 domain into pET-45b(+)-EGFP after the EGFP gene.

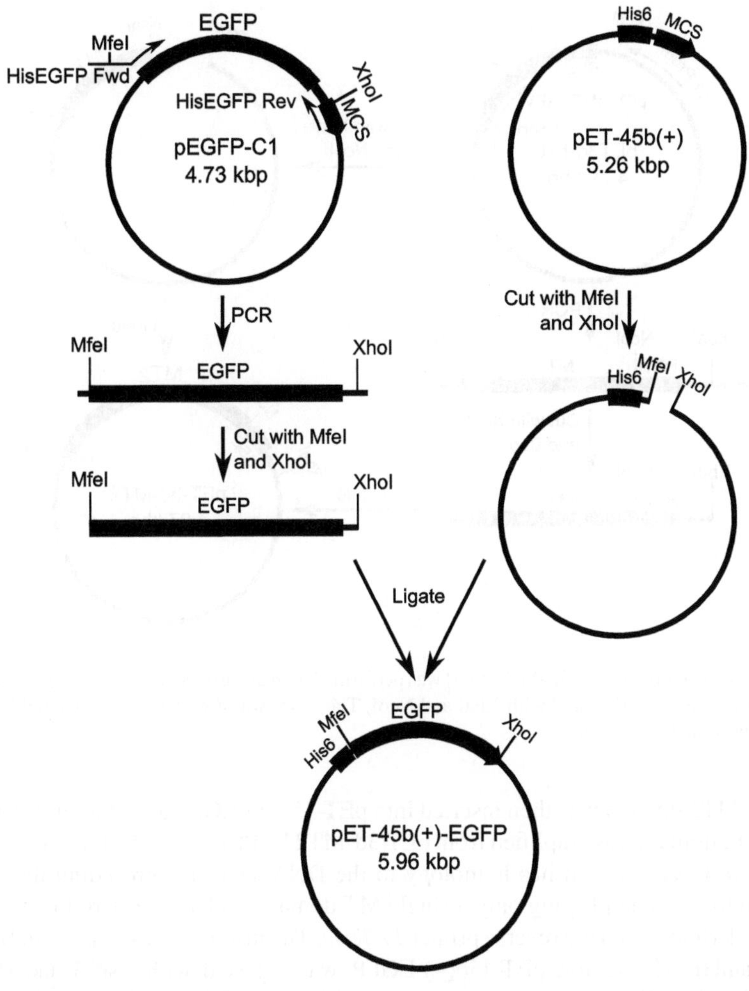

Fig. 2.1 Production of pET-45b(+)-EGFP, encoding a His_6-tagged EGFP fusion protein. PCR was performed with the primers *HisEGFP Fwd* and *HisEGFP Rev*. Both vector and insert were digested with MfeI and XhoI

The mouse MT-1 gene was amplified from pET-3d-MT (a gift from Prof. Winge, University of Utah) by gradient PCR using the forward primer *T7 Prom*, which retains the ribosome binding site, and the reverse primer *pET-3d-MT2 REV*, which adds a BspHI restriction site (underlined) and removes the MT stop codon. The insert was digested with XbaI and BspHI. The vector, pET-3d-MT, was digested with XbaI and NcoI. BspHI and NcoI generate complementary overhangs and ligate with the removal of both recognition sites. The insert was subcloned before the existing MT gene with no linker between the two domains to form pET-3d-MT2 (Fig. 2.2). This procedure was repeated to produce higher-order concatenations as necessary.

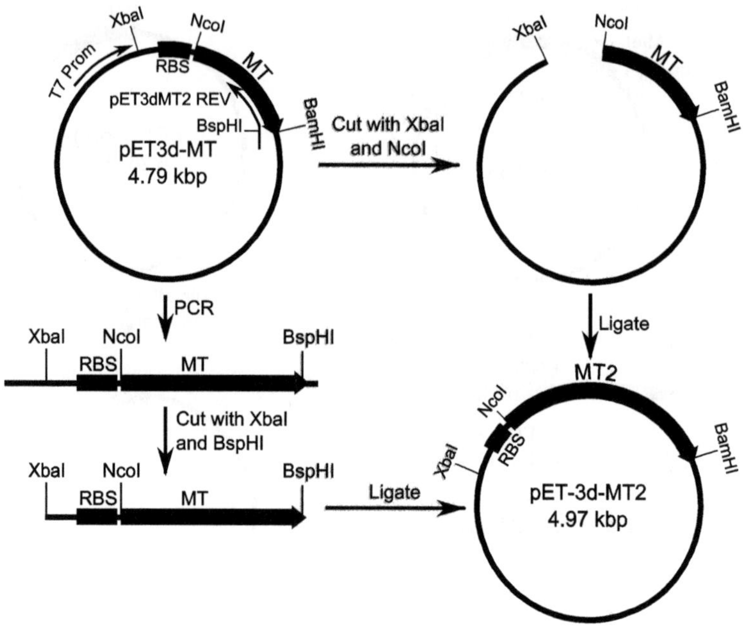

Fig. 2.2 Production of pET-3d-MT2. PCR was performed with the primers *T7 Prom* and *pET3dMT2 REV*. The vector was digested with XbaI and NcoI. The insert was digested with XbaI and BspHI. RBS; ribosome binding site

The MT2 domain was then inserted into pET-45b(+)-EGFP after the EGFP gene. The MT2 domain was amplified from pET-3d-MT2 by PCR using the forward primer *pET45-link-MT2*, which has homology to the DNA sequence preceding the MT2 domain to avoid amplifying only a single MT domain, and adds a BsrGI restriction site (underlined), and the reverse primer *T7 Term*. The insert was digested with BsrGI and BamHI. The vector, pET-45b(+)-EGFP, was digested with BsrGI and BglII. BamHI and BglII generate complementary overhangs and ligate with the removal of both recognition sites, creating pET-45b(+)-EGFP-MT2-C1 (Fig. 2.3).

2.3.1.3 His_6-Tagged EGFP-MT2-SNX1

Purifiable EGFP-MT2-SNX1 (EMS1) for membrane tubulation was achieved by cloning MT2-SNX1 into pET-45b(+)-EGFP to create pET-EGFP-MT2-SNX1 (pET-EMS1). This was achieved by first forming pEGFP-BsiWI-MT2-SNX1 by digesting pEGFP-MT2-C1 and pEGFP-SNX1 (for eukaryotic plasmids please see below) with AgeI and BsrGI prior to gel purification and ligation. This was then digested with BsiWI and SalI to yield sticky ended MT2-SNX1 whilst pET-45b(+)-EGFP was digested with BsrGI and XhoI. These were ligated without purification to yield pET-EMS1.

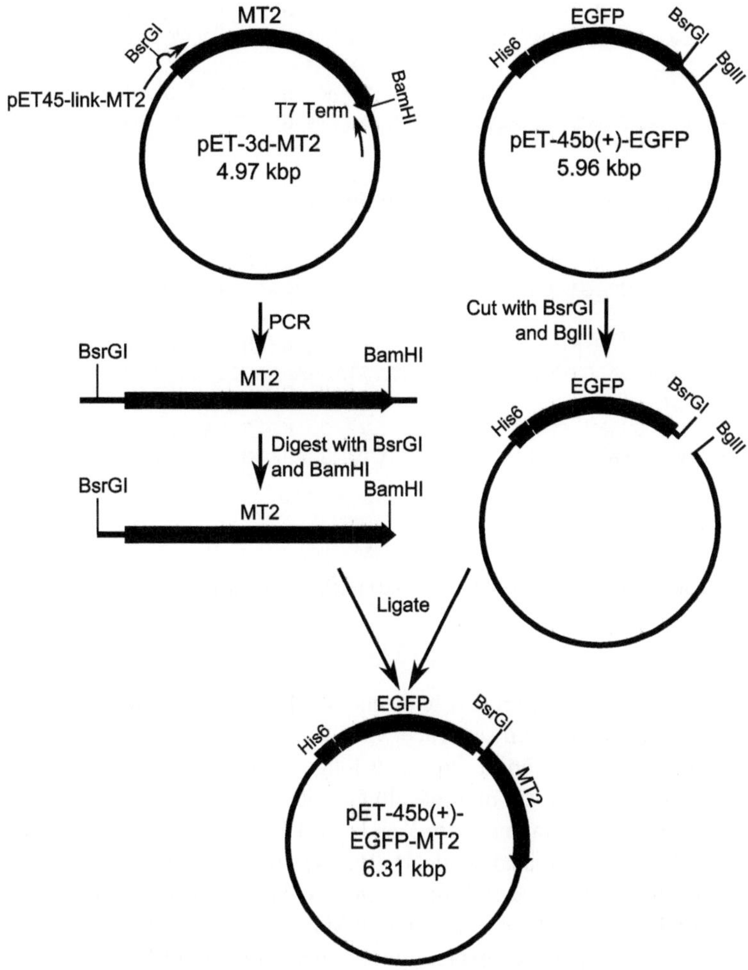

Fig. 2.3 Production of pET-45b(+)-EGFP-MT2, encoding a His_6-tagged EGFP-MT2 fusion protein. PCR was performed with the primers *pET45-link-MT2* and *T7 Term*. The vector was digested with BsrGI and BglII. The insert was digested with BsrGI and BamHI

2.3.2 Design and Synthesis of an MT2 Domain and pEGFP-MT2-C1

We required a concatenated MT domain to increase clustering of heavy-metal ions to form electron-dense nanoparticles. However a tandem gene fusion containing two identical sequences is undesirable due to potential homologous recombination and secondary structure elements interfering with PCR and expression. For this reason, a non-identical MT2 domain was designed using Gene Designer (DNA2.0, inc., California) and synthesised by GENEART (Regensburg, Germany) for insertion into

```
  1   M  D  P  N  C  S  C  S  T  G  G  S  C  T  C  T  S  S  C  A
  1   ATGGACCCTAATTGTTCCTGTTCTACTGGTGGAGTTGCACTTGTACCTCAAGCTGTGCT

 21   C  K  N  C  K  C  T  S  C  K  K  S  C  C  S  C  C  P  V  G
 61   TGTAAAAACTGCAAGTGCACGTCTTGTAAGAAGAGCTGTTGCTCATGTTGCCCAGTAGGG

 41   C  S  K  C  A  Q  G  C  V  C  K  G  A  A  D  K  C  T  C  C
121   TGTAGCAAATGCGCCCAGGGGTGCGTATGTAAAGGCGCAGCCGACAAATGTACCTGCTGT

 61   V  M  D  P  N  C  S  C  S  T  G  G  S  C  T  C  T  S  S  C
181   GTCATGGACCCGAACTGTAGCTGTAGCACTGGCGGGAGCTGTACTTGCACCAGTTCCTGT

 81   A  C  K  N  C  K  C  T  S  C  K  K  S  C  C  S  C  C  P  V
241   GCCTGCAAGAACTGTAAGTGCACATCATGTAAGAAAAGCTGCTGTTCTTGCTGTCCCGTG

101   G  C  S  K  C  A  Q  G  C  V  C  K  G  A  A  D  K  C  T  C
301   GGTTGCAGTAAATGCGCACAGGGCTGTGTATGCAAAGGGGCTGCCGATAAGTGCACCTGT

121   C  A
361   TGCGCA
```

Fig. 2.4 Sequence of the synthetic MT2 gene and protein. Amino acids are above the corresponding DNA codons

pEGFP-C1 (Fig. 2.4). The protein sequence of MT-1 was placed head-to-tail with no linker spacer. The protein sequence was then back-translated to DNA with the following restrictions; all repeat sequences longer than 10 bp were removed, all RE sites also found in pEGFP-C1 multiple cloning site (MCS) were removed, the BsrGI site found in pEGFP-C1 was moved to 3′ end of MT2 but kept in same reading frame, a BsiWI (which has an overhang identical to BsrGI) site was added at 5′ end of the MT2 domain (in same frame as BsrGI, for cutting out MT2 domain or swapping EGFP for EYFP, ECFP, etc), and the stop codon of MT was removed. Codon optimization was performed for human cell expression. For insertion into pEGFP-C1, a BbsI site was also added to the 5′ end of the insert that produces a 4 bp overhang downstream of the recognition site which was designed to be complimentary to that produced by BsrGI, but forms no recognition site for any restriction endonuclease. The fragment was cloned into pEGFP-C1, digesting the insert with BbsI and BsrGI and the vector with BsrGI with dephosphorylation, concomitantly removing the BbsI site whilst retaining BsrGI (Fig. 2.5).

2.3.3 *pEGFP-MT2-Tagged Proteins*

For eukaryotic expression, EGFP-MT2 was fused to various proteins that localise to specific regions of the cell within expression vectors. In each case, expression was under the control of the human cytomegalovirus (CMV) immediate early promoter.

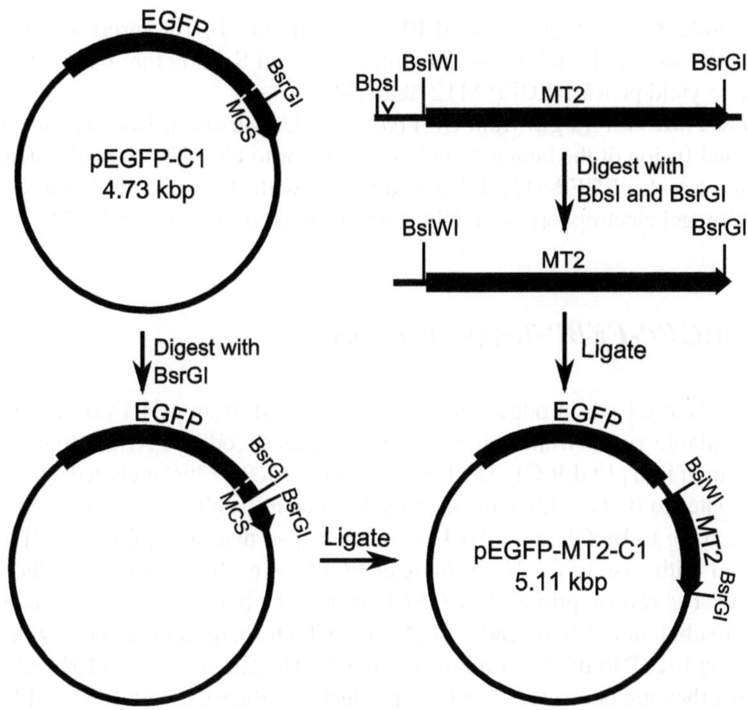

Fig. 2.5 Subcloning strategy to produce pEGFP-MT2-C1. The synthetic MT2 domain was digested with BsrGI and BbsI to leave a complementary overhang to BsrGI. The vector was digested with BsrGI and dephosphorylated before ligation of the fragments

pEGFP-Sec23A (a gift from Dr. David Stephens, University of Bristol) encodes a COPII coat component Sec23A fused to EGFP. pEGFP-Sec23A and pEGFP-MT2-C1 were digested with AgeI and BsrGI prior to purification by gel electrophoresis. The fragments were ligated as described above to form pEGFP-MT2-Sec23A.

pEGFP-SNX1 (a gift from Dr. Jan van Weering, University of Bristol) encodes sorting-nexin 1 (SNX1), which tubulates early endosomal compartments, fused to EGFP. pEGFP-SNX1 was digested with BglII and BsrGI whilst pEGFP-MT2-C1 was digested with BglII at 37 °C for 1 h, before BsiWI was added for a further incubation at 55 °C for 1 h. The fragments were isolated by gel purification prior to ligation to form pEGFP-MT2-SNX1.

pEGFP-tub (Clontech), encoding a fusion protein of EGFP and human α-tubulin, and pEGFP-MT2-C1 were digested with BsrGI and NheI prior to purification by gel electrophoresis and ligation to form pEGFP-MT2-Tubulin.

pCR3.1-Tetherin-GFP (a gift from Dr. George Banting and Mr. Pete Billcliff) encodes human Tetherin HuCD317 with GFP subcloned between bases 246 and 247 (after the glycosylation site and before the coiled-coil region of the protein). EGFP-MT2 was amplified from pEGFP-MT2-C1 using the primers *TethGFP For* and *TethMT2 Rev*, adding a 3′ SacII RE site (underlined). The insert was digested with BsiWI and SacII and the vector with BsrGI and SacII before purification and ligation to yield pCR3.1-EGFP-MT2-Tetherin.

pLVX-TfnR-GFP (a gift from Dr. Fred Boal, University of Bristol) contains an N-terminal fusion of the human transferrin receptor to GFP. This was digested with AgeI and SmaI, pEGFP-MT2-C1 was digested with AgeI and AfeI. These were purified by gel electrophoresis and ligated as described to form pEGFP-MT2-TfnR.

2.3.4 pEGFP-FKBP-Tagged Proteins

pEGFP-FKBP-C1 was produced by amplifying FKBP from pC4-Fv1E, a commercially available vector from Ariad Pharmaceuticals encoding FKBP12(F36V), and subcloning it into pEGFP-C1. The forward primer *FKBP FWD* included a BbsI site, which ligates to BsrGI with removal of the RE site, a BsiWI site, which has an identical overhang to BsrGI, and a SpeI site, included so that, if required, FKBP could be moved with ease to the N-terminus of EGFP using NheI, which has the same overhang. The reverse primer, *FKBP REV*, included a BsrGI site to maintain the RE site after subcloning FKBP, and a SgrAI site, which shares an overhang with AgeI for moving FKBP to the N-terminus if required. The RE sites are underlined in the order that they are described. The PCR product was digested with BbsI and BsrGI, whilst pEGFP-C1 was digested with BsrGI and dephosphorylated with calf intestinal alkaline phosphatase. The digested vector was purified and FKBP ligated in place.

pEGFP-FKBP-tagged variants of the genes described above for MT2-tagging were produced by digesting out EGFP-MT2 with AgeI and BsrGI and subcloning in EGFP-FKBP with the same REs.

2.4 Protein Expression and Purification

2.4.1 Inducing Protein Expression

pET-45b(+) contains the T7*lac* promoter which requires IPTG for high-level expression. However, in BL21(DE3) cells lacking pLysS there is a low level of background expression. Selective LB was inoculated with 1:40 dilution of overnight culture and grown for 1 h with shaking at 250 rpm.

For expression of His_6-EGFP, IPTG was added to a final concentration of 1 mM prior to incubation at 37 °C for 5 h. Due to the high cysteine content of His_6-EGFP-MT2 high-level expression was undesirable, so after dilution of overnight culture 0.05 mM IPTG was added, followed 30 min later by 0.3 mM $ZnCl_2$. Cells were then incubated overnight at 20–25 °C. pET-EMS1 expression appeared to be more stable than EGFP-MT2 expression, therefore 1 mM IPTG was used, followed by

0.3 mM $ZnCl_2$ as before. Expression was continued for 5 h. Cells were harvested by centrifugation at 8000 g for 10 mins.

2.4.2 *Purification of His_6-Tagged Protein*

His_6-tagged protein was purified using nickel(II)-nitrilotriacetic acid (Ni-NTA)-agarose beads (QIAGEN). Lysis, wash and elution buffer compositions were as described in the QIAexpressionist handbook (QIAGEN).

Purification of His_6-EGFP, EGFP-MT2 and EMS1 followed standard procedures. Briefly, the cell pellet was resuspended in lysis buffer and incubated with lysozyme at 1 mg/ml on ice for 30 min before freeze-thaw twice in liquid nitrogen. The solution was then sonicated for three ten-second bursts, all performed on ice. Clarification by centrifugation gave a bright green supernatant that was then incubated at 4 °C with Ni-NTA resin, with rocking, for 30 mins. The sample was loaded into a column with a frit (20 μm pores) and subject to 2 column washes with wash buffer. Finally, purified protein was eluted with high imidazole concentration. Buffer exchange was performed by concentrating protein in Amicon Ultra-30K spin concentrators (Millipore), resuspending in 25 mM Tris·HCl pH 8, and repeating twice more.

2.4.3 *Polyacrylamide Gel Electrophoresis (PAGE)*

Samples were run on 4–12 % Bis-Tris polyacrylamide gels at 150 V for 90 min and stained with Coomasie Blue R250 for 10 min. The gel was washed twice with destaining solution, each for 1 hr, prior to a final overnight wash in deionised H_2O.

2.5 Metal-Binding Assays

2.5.1 *Mass Spectrometry*

Matrix-assisted laser desorption/ionization mass spectrometry (MALDI MS) was used to elucidate the mass-to-charge ratio of purified EGFP-MT2 before and after incubation with heavy-metal salts, and therefore calculate the number of ions bound to the protein. Buffers tested were 0.1 M ammonium acetate, 0.1 M phosphate buffer, PBS, or 25 mM Tris·HCl, all at pH 8. Matrixes tested were 2,5-dihydroxybenzoic acid ±0.1 % formic acid or ±50 % acetonitrile, α-cyano-4-hydroxycinnamic acid, sinapinic acid and 6-aza-thiothymine. Each with or without β mercaptoethanol or TCEP, both at either 1 or 20 mM. This was attempted for each metal complex at cysteine:metal ratios of 10:1, 1:1, 1:10 after 3 h at 37 °C (50 μM protein contains 40 cysteine residues and is equivalent to 2 mM cysteine). The signal-to-noise ratio varied widely depending on which matrix and buffer was used, as well the concentration of protein and metal. Tris·HCl was optimal, combined with 2,5-dihydroxybenzoic acid (DHB) with 0.1 % formic acid as a matrix. Serial dilutions of protein in matrix were spotted onto a MALDI plate and analyzed using linear-mode on a MALDI 4700 Proteomics Analyzer (Applied Biosystems, California).

2.5.2 *Synthesis of Gold Nanoparticles*

Gold nanoparticles (AuNPs) were synthesised as described by Haiss et al. [1]. Briefly, a 1 % chloroauric acid ($HAuCl_4 \cdot 3H_2O$, w/v) was prepared in deionised water. 90 ml of deionised water was rapidly stirred and 1 ml chloroauric acid added. After 1 min of continuous stirring, 2 ml of 38.8 mM trisodium citrate was injected and stirred for a further minute. Finally, 1 ml of 0.075 % $NaBH_4$ (w/v) in 38.8 mM trisodium citrate was added and stirred for 5–10 min. The concentration of AuNPs was determined by UV-vis spectrometry as described in [1] and the size determined by DLS and TEM.

2.5.3 *Large Unilamellar Vesicles*

Vesicles were prepared by sonication and extrusion as described [2]. Briefly, stock palmitoyloleoyl phosphatidylcholine (POPC) lipids were diluted to 2.5 mg/mL in chloroform:methanol (10:1 v/v) in glass vials and dried under gentle nitrogen current. All traces of solvent were removed by freeze-drying for at least 2 h. The lipids were rehydrated using 1–1.5 mL of protein solution or buffer prior to sonication for 30 min at 40–50 °C. Finally, the solution was extruded 11 times through a polycarbonate membrane perforated with 200 nm pores. Diameter and size distribution of vesicles was confirmed by DLS.

Vesicles were incubated with metal compounds at room temperature for 3 h. For imaging in the TEM, vesicles were either placed directly onto formvar and carbon-coated grids and allowed to air-dry, or incubated with methylcellulose to stabilise the lipids. To coat with methylcellulose, vesicles were pipetted onto formvar and carbon-coated grids and allowed to settle for 5 min. Three washes with deionised water was followed by addition of 1.8 % methylcellulose and 0.3 % uranyl acetate in deionised water on ice. After 5–10 min grids were looped out and excess fluid was wicked away.

2.6 Membrane Tubulation

Membrane tubulation by SNX-BAR proteins requires the presence of phosphatidylinositol-3-phosphate (PI(3)P) as well as other ancillary phospholipids to create microdomains suitable for tubulation. Liposomes were formed at 1 mg/mL from bovine brain extract (Folch fraction 1, Sigma-Aldrich and Avanti batches were mixed 1:1) dissolved in chloroform and supplemented with 4 % protonated PI(3)P (Avanti polar lipids). PI(3)P was protonated by dissolving the non-protonated lipids in chloroform before removing the solvent under nitrogen and desiccation for 1 h. The lipid film was redissolved in chloroform:methanol:1N HCl (2:1:0.01 v/v) and left for 15 min. These were dried and desiccated as before and washed with chloroform:methanol (3:1 v/v) and twice with chloroform.

Lipid solutions were treated as above and extruded to either 400 or 200 nm in diameter. Purified pET-EMS1 was incubated with AuTM (1:1 gold to cysteine residues) before being added to the liposomes at 17.5 μM, incubated at room temperature for 10 min and spotted onto carbon-coated EM grids. The liposomes were stabilised with methylcellulose and either left unstained or negatively stained with uranyl acetate before viewing in the TEM.

2.7 SAF Formation

Peptides were synthesised by solid-phase peptide synthesis by Dr. Thomson, University of Bristol, and purified by high-performance liquid chromatography. The sequence of each peptide is shown in Table 2.1. Freeze-dried peptides were rehydrated in deionised water to form stock concentrations at ~1 M. Stocks were diluted to a final concentration of 0.1 mM in 10 mM MOPS pH 7.4 and grown for 18 h at 21 °C.

Table 2.1 SAF peptide sequences. Sequences start at g and end at f because g–e interactions are between successive heptads

Peptide	Sequence
Heptad repeat	g abcdefg abcdefg abcdefg abcdefg abcdefg abcdef
SAF-p1	K IAALKQK IASLKQE IDALEYE NDALEQ
SAF-p1-I	K IAALKQK IASLKQE IDALEFE NDALEQ
Blunt-p1-I	E IDALEFE NDALEQK IAALKQK IASLKQ
SAF-p2a	K IRRLKQK NARLKQE IAALEYE IAALEQ
Blunt-p2a-I	E IAALEFE IAALEQK IRRLKQK NARLKQ
p1$_3$-N	Ac-K IASLKQE IDALEYE NDALEQ
p1$_3$-C	K IAALKQK IASLKQE IDALEY-Am
p2a$_6$	Ac-E IAALEQK IRRLKQK NARLKQE IAALEYE IAALEQK IRRLKQ-Am

F represents the *p*-iodo-phenylalanine. Ac and Am represent acetylated and amidated termini, respectively

References

1. W. Haiss , N.T.K. Thanh , J. Aveyard , D.G. Fernig, Determination of size and concentration of gold nanoparticles from UV-Vis spectra. Anal. Chem. **79**(11), 4215–4221 (2007)
2. B.L. Mui, P.R. Cullis, E.A. Evans, T.D. Madden, Osmotic properties of large unilamellar vesicles prepared by extrusion. Biophys. J. **64**(2), 443–453 (1993)

Chapter 3
Imaging Methods

3.1 Light Microscopy

3.1.1 Microscopes

For widefield microscopy, a Leica DMIRB inverted microscope was used for fluorescence and phase contrast imaging of live *E. coli* cells. For fluorescence imaging of fixed cells, an Olympus IX71 inverted microscope with an Exfo Excite illumination system and Ludl filter wheel were used with a Photometrics Coolsnap HQ^2 CCD camera. Both microscopes use Improvision Volocity software for acquisition of images.

For confocal microscopy, a Leica confocal imaging spectrophotometer system (TCS-SP2) was used for phase contrast and fluorescence imaging of live cells. Proprietary Leica software was used for image acquisition.

A spinning disk microscope was used for high-resolution time-lapse imaging. A Perkin Elmer spinning disk system attached to a Leica DMI6000 was used with an argon laser at 488 nm. Images were captured on a Hamamatsu EMCCD camera controlled by Improvision Volocity software.

3.1.2 Transformed Escherichia coli

For fluorescent imaging of bacterial cells, 10 μl of bacterial culture was pipetted onto a microscope slide and the coverslip was held on with tape.

T. H. Sharp, *Biomolecular Imaging at High Spatial and Temporal Resolution In Vitro and In Vivo*, Springer Theses, DOI: 10.1007/978-3-319-02159-1_3,

3.1.3 *Transfected HeLa Cells*

3.1.3.1 Immunofluorescence

Cells were grown to approximately 50 % confluency on 22 mm glass coverslips and transfected with pEGFP-MT2-Sec23A. After an overnight incubation, cells were fixed by methanol fixation. Cells were washed with PBS and fixed at −20 °C in methanol (already at −20 °C) for 4 min. Coverslips were then washed three times with PBS to remove traces of methanol. Non-specific antibody binding was blocked with 3 % BSA in PBS (BSA-PBS) for 30 min. All antibody binding was performed at room temperature in a humidified atmosphere. Primary antibodies were diluted into BSA-PBS as described in Table 3.1 and incubated for 1 h. Cells were washed three times for 10 min with PBS before addition of secondary antibodies, diluted 1:400 in BSA-PBS, for a further 1 h. Three further washes with PBS were performed before counter-staining with 4′,6-diamidino-2-phenylindole (DAPI) diluted 1:5000 in PBS. Coverslips were mounted with Mowiol 4-88 (Polysciences, Inc.) and stored in foil at 4 °C.

3.1.3.2 Live-Cell CLEM

Cells were grown to 50 % confluency in 35 mm dishes containing a CelLocate coverslip glued in place for chemical fixation. After an initial imaging to confirm transfection, cells were incubated in DMEM in the presence of either 0.1 mM AuTM, 0.1 mM $KAuCl_4$ or 10 μM $CdCl_2$ at 37 °C for 3 h. Cells were then imaged with a 63× oil-immersion objective lens. First, an overview of the locator grid squares was imaged to enable localization of the region of interest during sectioning before the cell of interest was imaged at higher magnification. Cells were fixed with glutaraldehyde as described below.

Table 3.1 Primary and secondary antibody dilutions

Primary antibody	Dilution (species)	Secondary antibody
Anti-Sec24C	1:100 (R)	Cy3-anti-rabbit
Anti-Sec31A	1:100 (R)	Cy3-anti-rabbit
Anti-Giantin	1:2000 (R)	Cy3-anti-rabbit
Anti-GM130	1:1000 (M)	Cy3-anti-mouse
Anti-β'-COP	1:100 (R)	Cy3-anti-rabbit
Anti-Calnexin	1:500 (R)	Cy3-anti-rabbit

R and M denote rabbit and mouse, respectively

3.2 Electron Microscopy

3.2.1 Microscopes

Two TEMs were used in this study. A Tecnai T12 (FEI company, Eindhoven, The Netherlands) with a biotwin objective lens operating at 120 kV was used to image all negatively stained specimens, whereas a Tecnai T20 with a LaB_6 filament operating at 200 kV was used to perform cryoTEM, STEM, and electron tomography.

3.2.2 Chemical Fixation

Bacterial samples were pelleted by centrifugation and most of the supernatant removed. The pellet was submerged in 2.5 % glutaraldehyde (GA) in 0.1 M Cacodylate buffer for 20 min before resuspension. Cells were re-pelleted and submerged in 1 % OsO_4 in deionised water, gently resuspended, immediately re-pelleted and then left for 20 min. The supernatant was removed and the pellet washed three times with deionised water. Dehydration with increasing concentrations of ethanol in deionised water was performed before Epon infiltration. After sectioning, samples were transferred onto a carbon-coated grid and stained with 1 % uranyl acetate.

HeLa cells were fixed with 2.5 % GA in 0.1 M Cacodylate buffer for 1–2 h at room temperature. After washing with 0.1 M Cacodylate buffer cells were post-fixed with 0.1 M osmium tetroxide for 30 min at room temperature, washed extensively with deionised water before dehydration in ethanol (70–100 %). Epon was added and after 2 h replaced with fresh Epon. Finally, a stub was placed over the cells and cured for 2 days at 60 °C.

3.2.3 Gold Enhancement

Gold enhancement was performed on whole cells directly after chemical fixation. GoldEnhance EM (Nanoprobes, USA) was used with a modified protocol as described by Morphew et al. [1, 2]. Briefly, aldehyde groups were quenched with PBS-Gly and washed with PBS-BSA prior to a final wash in Solution E (see Appendix B). The manufacturer's solution A was mixed with solution B (2:1) and added to the cells. After 5 min, solutions E and D were added (5:2) and developed at 4 °C for 10 min. The reaction was halted by adding 1.5 % sodium thiosulphate for 5 min and washed extensively with solution E. Immunolabelling or osmification proceeded as described.

3.2.4 *Tokuyasu Cryosectioning*

HeLa cells were seeded in 10 cm cell culture dishes and grown to 80 % confluency. Fixation was initiated with 2 % paraformaldehyde (PFA) with 0.025 % GA in 0.1 mM PB, pH 7.4 for 1 h at room temperature. The fixative was exchanged with 4 % PFA with 0.05 % GA and continued overnight at 4 °C. Cells were washed three times with PB and soaked in 1 % fish gelatine. A soft teflon block was used to gently scrape the cells into a 1.5 ml Eppendorf tube before they were pelleted for 5 s at 4000 g. The supernatant was removed before the cells were resuspended in 12 % gelatine and allowed to infiltrate for 10 min at 37 °C. The cells were pelleted once more and the gelatine set by incubation on ice for 1 h. The bottom of the tube containing the pelleted cells was removed with a razor blade and the pellet cut into 1 mm cubes. The cubes were incubated in 2.3 M sucrose at 4 °C overnight. Cubes were placed on aluminium rivets and frozen in LN2. Blocks were trimmed at −80 °C and sectioned at −120 °C. Sections were picked up in 1:1 2 % methylcellulose:2.3 M sucrose and counterstained with 0.3 % uranyl acetate in 1.8 % methylcellulose or immunolabelled as described below.

3.2.4.1 Immunolabelling

Immunolabelling was performed on either cryosectioned or chemically-fixed whole cells. Immunolabelling of cryosections was performed immediately after section retrieval by floating the grids on PBS-Gly then PBS-BSA before labelling with Rabbit-Anti-GFP (A11122, Invitrogen) dissolved 1:25 in PBS-BSA for 20 min. Protein-A conjugated to 10 nm gold (Aurion) diluted 1:10 in PBS-BSA was incubated at room temperature for 20 min and washed with PBS. Counterstaining was performed as described above.

To immunolabel whole cells, fixation with 4 % PFA was followed by extensive washing with PBS-Gly and blocking in PBS-BSA. Rabbit-anti-tetherin (a gift from Mr. Pete Billcliff) was diluted 1:300 in PBS-BSA and incubated with the cells for 1 h in a humidified atmosphere. After further washing, goat-anti-rabbit-10 nm gold (Aurion) was incubated with the cells for a further hour. This followed by post-fixation with 2.5 % GA before osmification and dehydration.

3.2.5 *Processing Grids for CLEM*

During live-cell imaging, the CelLocate grid was imaged for later relocation of cells (Fig. 3.1a–c). After processing as described above Epon resin stubs were removed from the glass coverslips by repeatedly cooling the glass in LN2 and rapidly heating it in boiling water. The CelLocate grid was clearly visible under a stereomicroscope, which enabled tracing back of the same set of cells (Fig. 3.1d). The resin was trimmed

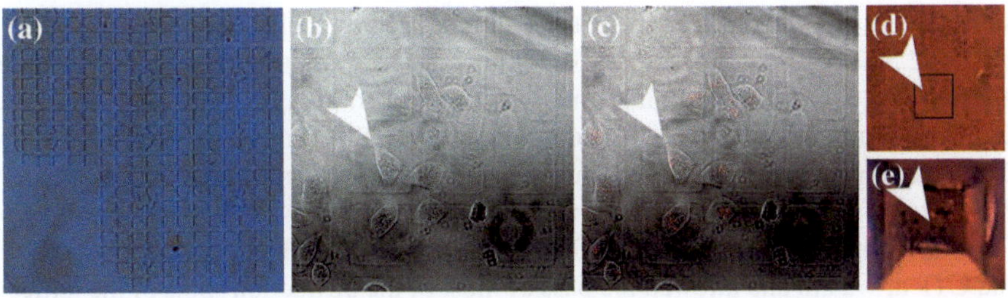

Fig. 3.1 Preparation of EM grids for sectioning. **a–c** A cell of interest is imaged by FLM and phase contrast microscopy, along with an overview of the finder grid. The sample is fixed and embedded as described above. **d** and **e** The same region is traced back and trimmed to a *trapezium* around the area of interest. Modified from [3]

with a razor blade around the grid square imaged in the fluorescence microscope overview image before trimming into a trapezium immediately surrounding the area of interest using a fine razor blade (Fig. 3.1e). The stubs were then sectioned to give serial thin (70 nm) or semi-thin (300 nm) sections, which were collected on slot grids.

3.2.6 Negative Stain

SAF samples were prepared as described above before 5 μl sample was placed on carbon-coated copper grids. After 1 min the sample was wicked off and stained with 10 μl of 3 % uranyl acetate for 30 s. The stain was wicked off and the grid air-dried for 10 min before insertion into the microscope.

3.2.7 Electron Tomography

Carbon-coated copper grids were incubated with 10 nm colloidal gold (Aurion, NL) on both sides to act as fiducial markers. Samples were prepared by negative staining as described. Tomographic tilt series were collected semi-automatically using FEI software (FEI Company, NL). Images were saved as .tif stacks, converted using tif2mrc and processed using IMOD [4] running under Cygwin (http://cygwin.com).

3.3 Electron Cryo-Microscopy

SAF samples were prepared as described above. Lacey-carbon grids (Electron Microscopy Sciences, PA, USA) were glow discharged before insertion into the Vitrobot (FEI Company). Humidity was set to 100 % before addition of 5 μl of sample. For high-resolution calibration, samples were incubated with T4 bacteriophage; 2.5 μl SAF solution was mixed with 2.5 μl T4 solution immediately before adding 5 μl to the grid. Blotting time was 1 s prior to plunging into liquid ethane. Grids were transferred to a cryo-holder (Gatan, Inc, USA) and visualised in a Tecnai T20 electron microscope (FEI Company) at 200 kV accelerating voltage at a magnification of ×50 k with an electron dose of 10 electrons $Å^{-2}$. Low dose software (FEI Company) was used with a defocus of −1.5 μm, which placed the first zero of the CTF at ~2 nm and allowed the pass of most of the diffraction spots of the lattice.

3.3.1 Electron Cryo-Tomography

Freshly glow-discharged lacey-carbon grids (Electron Microscopy Sciences) were incubated with 10 nm colloidal gold (Aurion) on both sides to act as fiducial markers. Samples were prepared by plunge freezing as described. Processing of tilt series was as described above.

3.4 Image Processing

3.4.1 Electron Crystallography

Images were analyzed using a combination of the programs ImageJ (NIH, http://rsbweb.nih.gov/ij/) and 2dx [5]. Subsidiary programs found in Collaborative Computational Project No. 4 (CCP4) (http://www.ccp4.ac.uk/) [6] were also used by 2dx but not accessed directly. Axes and lattice plane nomenclature are as shown in Fig. 3.2. First, ImageJ was used to rotate images so that the fibre long (z-)axis was vertical (as judged from the layer lines of the FT of the image) and the image was cropped to a square image containing the feature of interest. 2dx_Image was then used for all further processing of individual images. First some parameters were set as universal for the processing of all images based on the high-resolution calibration by T4 bacteriophage and the average values of the image data; the pixel size was 2.14 Å, the real-space lattice was set as 18 Å × 125.4 Å and the angle between them as 90 °. The C_s and accelerating voltage were 2 mm and 200 kV, respectively. The upper resolution of processing was set to 5.5 Å. The tilt of the specimen was set to zero due to the assumption that any super-helical pitch of the fibre with respect to the coiled coils was independent of sample tilt. Regions of the SAFs were then masked from the rest of the image with an edge taper to reduce FT artifacts. The FT of the image was calculated and the reciprocal-space lattice defined. The CTF was automatically refined from 1.5 μm. Unbending parameters were modified so that a smaller box size than the default was used due to the small unit cell dimensions and high regularity of the lattice. Reflections were identified automatically and refined

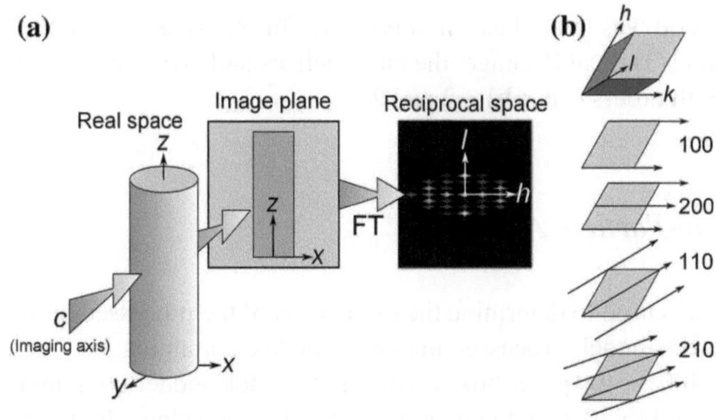

Fig. 3.2 Axes and lattice plane nomenclature. **a** Axes labeled as they appear in the text in real space, on the image plane, and in reciprocal space. **b** Space group nomenclature is standard (*hkl*). The 4 *hexagonal lattice planes* discussed in this Thesis are shown for reference

manually if required. Unbending was repeated twice prior to CTF correction. Finally, 2dx_generateMap was invoked to produce a greyscale image and contour plot of 2×2 unit cells.

2dx_Merge was used to attempt to merge the data to reduce the noise and increase the resolution of the density maps. Since the SAFs could be in any one of four orientations (see Sect. 4.2) two methods were attempted. Firstly, the phase origin of each image was set manually and refined automatically by searching over 6° in 0.1° steps. The images were either rotated 180° around the c-axis (imaging axis, i.e. $h = -h$, $l = -l$) using ROT180, mirrored by rotating 180° around the z-axis (parallel to the fibre long axis, i.e. $h = -h$) using REVXSGN, or both (i.e. $l = -l$). The preferred orientation was the one with the lowest phase residuals. Secondly, the images were aligned manually and refined against the model as a reference. Successive rounds of testing whether the images should be rotated or mirrored were followed with rounds of refining and merging, either against the model data or against the manually aligned images. However, since it was not possible to automatically determine the orientation of each fibre, merging of the data was not viable. Therefore, the best images (those with clear reflections and phases out to the highest resolution) were chosen for model refinement.

3.4.2 Helical Reconstruction

ImageJ was used to create a single coiled coil 10 pitches long from a CTF-corrected map generated by 2dx, which was floated before further processing with the Brandeis Helical Package [7]. Prior knowledge of coiled coil geometry was used to elucidate the Bessel orders, as described in the main text (see Sect. 4.4). At the working resolution of 6–10 Å the difference between the heterodimeric α-helices cannot be identified so the coiled coil was treated as a homodimer. This restricts n to multiples of 2 [8]. For the Brandeis Helical Package, the basic helices had layer line heights ($l\mathbf{a}$, $l\mathbf{b}$) of 10 and 20, with orders ($n\mathbf{a}$, $n\mathbf{b}$) −2 and 2.

3.4.3 Single-Particle Analysis

EMAN2 [9] was used to determine the orientation of the processed images and validate the all-atom model. Processed images from 2dx containing 5 × 2 unit cells were padded into 400 × 400 pixel boxes. An all-atom pdb model containing 19 coiled coils was filtered to 10 Å and converted to mrc format using e2pdb2mrc.py with a box size of 400 × 400 and pixel size of 0.627 Å to match the processed images. Projections of the model sampled C1 space in 5° increments. Using e2refine.py, micrographs were aligned over one iteration to the best-matching projection image using rotate_trans_flip_scale_iter. Aligned images were identified using e2display.py.

3.5 Model Building and Molecular Dynamics

All images of protein structures presented in this Thesis were generated with either PyMol or UCSF Chimera.

The radius (r) of the coiled coils present in SAFs, defined as the radius of the center of the α-helices from the center of the coiled coil, was measured from the images as 4.4 Å. Coiled coils were modelled with a rise per residue (t_{res}) of 1.529 Å and a $\phi_{C\beta}$ of 205°, as defined in [10]. The value for t_{res} was chosen so that equivalent residues had the same x, y coordinates after one pitch (125.4 Å in the z direction) and can be calculated using Eq. (3.1) [10, 11]:

$$t_{res} = \left(2 - \frac{7}{\tau}\right)\left(\frac{\sqrt{P^2 + 4\pi^2 r^2}}{7}\right) \tag{3.1}$$

Where τ is the residues per turn of an α-helix and;

$$\tau = \frac{m}{n} \tag{3.2}$$

Since the SAFs are a 3_1 helix, there are (3 × 28) 84 residues per coiled-coil pitch (m). There must also be an integral number of turns of each α-helix (n), which means, constrained by the requirement of ~3.6 residues per turn of α-helices in left-handed coiled coils, there must be 23 turns of each α-helix.

Sequences of p1 and p2a were repeated three times and modelled as continuous coiled coils using the program *makeccsc* (kindly provided by Dr. Gerald Offer, University of Bristol) with side chain coordinates taken from a database library of preferred conformations. The side chains experiencing intra-coiled coil interactions were then moved to the rotamers found in the crystal structure using InsightII (2005, Accelrys Inc.). The backbone was broken to simulate the sticky ended building blocks and the extra oxygen and hydrogen atoms added. Finally, the N- and C-termini were rotated to avoid steric clashes.

Using PyMol (DeLano Scientific, http://www.pymol.org/), the coiled coils were placed on a hexagonal lattice 20.8 Å apart. To model the effect of the MOPS buffer, three different scenarios were simulated: MOPS present next to R_4; MOPS present bound to R_{10}; and no MOPS present (Cl^- ions were added instead to maintain the charge balance). These were then subjected to molecular dynamics simulations (in collaboration with Dr. Richard Sessions, University of Bristol) or used directly to simulate electron density projections.

3.5.1 Model Fitting

The EM density map was thresholded to enclose the volume of either 100 or 60 % of the mass of the protein model. The values were calculated based on the weight of three heterodimers (19,496 Da; corresponding to a single pitch of the coiled coil) multiplied by 1.21 $Å^3 Da^{-1}$ [12, 13].

Atomic models were fit into the EM density map as rigid structures. UCSF Chimera [14] was used to first manually place the model within the EM density map before simulating the PDB model density to 8 Å and optimizing the correlation between the two maps. The fitting routine was repeated until convergence. Different starting orientations were also tried.

References

1. M. Morphew, W. He, P.J. Bjorkman, J.R. McIntosh, Silver enhancement of nanogold particles during freeze substitution for electron microscopy. J. Microsc. **230**(Pt 2), 263–267 (2008)
2. W. He, C. Kivork, S. MacHinani, M. Morphew, A. Gail, D. Tesar, N. Tiangco, J. McIntosh, P. Bjorkman, A freeze substitution fixation-based gold enlarging technique for EM studies of endocytosed nanogold-labeled molecules. J. Struct. Biol. **160**(1), 103–113 (2007)
3. J.R. van Weering, E. Brown, T.H. Sharp, J. Mantell, P.J. Cullen, P. Verkade, Intracellular membrane traffic at high resolution. Methods Cell Biol. **96**, 619–648 (2010)
4. J.R. Kremer, D.N. Mastronarde, J.R. McIntosh, Computer visualization of three-dimensional image data using IMOD. J. Struct. Biol. **116**(1), 71–76 (1996)
5. B. Gipson, X. Zeng, Z. Zhang, H. Stahlberg, 2dx-User-friendly image processing for 2D crystals. J. Struct. Biol. **157**(1), 64–72 (2007)
6. M.D. Winn, C.C. Ballard, K.D. Cowtan, E.J. Dodson, P. Emsley, P.R. Evans, R.M. Keegan, E.B. Krissinel, A.G.W. Leslie, A. McCoy, S.J. McNicholas, G.N. Murshudov, N.S. Pannu, E.A. Potterton, H.R. Powell, R.J. Read, A. Vagin, K.S. Wilson, Overview of the CCP4 suite and current developments. Acta Crystallogr. Sect. D **67**(4), 235–242 (2011)
7. C.H. Owen, D.G. Morgan, D.J. DeRosier, Image analysis of helical objects: the Brandeis helical package. J. Struct. Biol. **116**(1), 167–175 (1996)

8. M. Stewart, Computer image processing of electron micrographs of biological structures with helical symmetry. J. Electron Microsc. Tech. **9**(4), 325–358 (1988)
9. G. Tang, L. Peng, P. Baldwin, D. Mann, W. Jiang, I. Rees, S. Ludtke, EMAN2: an extensible image processing suite for electron microscopy. J. Struct. Biol. **157**(1), 38–46 (2007)
10. G. Offer, R. Sessions, Computer modelling of the alpha-helical coiled coil: packing of side-chains in the inner core. J. Mol. Biol. **249**(5), 967–987 (1995)
11. R.D.B. Fraser, T.P. MacRae, G.E. Rogers, in *Keratins: Their Composition, Structure, and Biosynthesis*, ed. by C.C. Thomas (Springfield, Illinois, 1972)
12. Y. Harpaz, M. Gerstein, C. Chothia, Volume changes on protein folding. Structure **2**(7), 641–649 (1994)
13. T. Goddard, C. Huang, T. Ferrin, Visualizing density maps with UCSF chimera. J. Struct. Biol. **157**(1), 281–287 (2007)
14. E.F. Pettersen, T.D. Goddard, C.C. Huang, G.S. Couch, D.M. Greenblatt, E.C. Meng, T.E. Ferrin, UCSF chimera-a visualization system for exploratory research and analysis. J. Comput. Chem. **25**(13), 1605–1612 (2004)

Part III
Experimental Results

Chapter 4
Towards an Atomistic Structure of a *De Novo* Designed Peptide Fibre

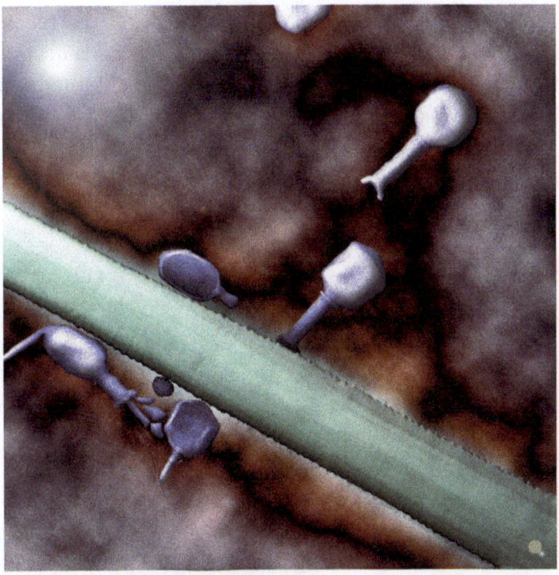

Recently, various applications for self-assembling peptide and protein fibres have come to the fore in biotechnology, one such system developed in the Woolfson lab is the self-assembling peptide fibre (SAF) system.

CryoTEM was used to image the superstructure of SAFs in a near-native hydrated state. This yielded a great deal of information regarding the packing regime of coiled coils within the fibre, described in Sect. 4.1, which is followed by a description of how this data was used to calculate a 6 Å electron density projection map of the composite coiled coils (Sect. 4.2). X-ray crystallography of circularly-permuted SAF peptides resulted in a 2.3 Å structure of the individual building blocks (Sect. 4.3). Next, in Sect. 4.4, a 3D electron density map was generated with a resolution of 8 Å, into

---

Results from this Chapter are published in [1].

T. H. Sharp, *Biomolecular Imaging at High Spatial and Temporal Resolution In Vitro and In Vivo*, Springer Theses, DOI: 10.1007/978-3-319-02159-1_4,

which an atomistic model of the composite coiled coils could be placed. These data were used to generate an all-atom model that was used to test the predictions made and identify the amino acids that allow the formation of the hexagonal array, described in Sect. 4.5. This chapter concludes with a discussion of the results presented herein, together with the implications for structure determination of intermediate filaments (Sect. 4.7).

4.1 CryoTEM of the SAFs

The SAFs display considerable superstructural order when imaged by negative stain TEM, manifest as lateral striations perpendicular to the long axis of the fibre, which extend across the entire width and along the length of each fibre with a periodicity of ∼42 Å [2] (Fig. 1.24). Using cryoTEM to image the SAFs revealed fibres with lateral striations with a similar repeat distance (Fig. 4.1a). To accurately determine the width of these striations the microscope was first calibrated with a sample of catalase crystal [3] before samples of fully formed SAFs were incubated with T4 bacteriophage. These have a well-defined striation along their tails of 40.6 Å [4] and as such were an ideal calibrant to image concomitantly with the ∼42 Å lateral SAF

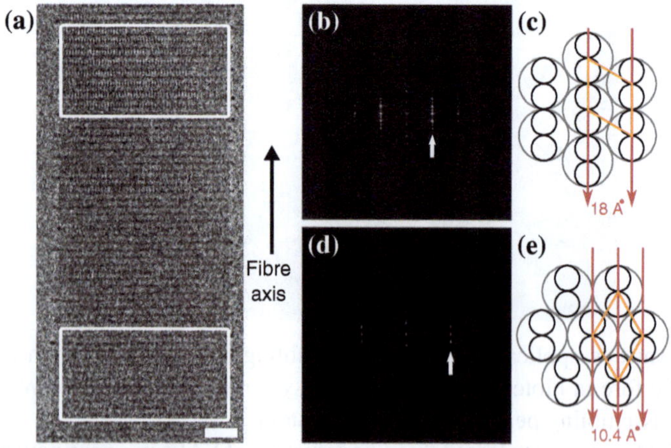

Fig. 4.1 Lateral and longitudinal striations present in frozen-hydrated SAFs. **a** Representative cryoTEM of a single frozen-hydrated SAF, revealing the lateral striations across the width and along the length of the entire fibre, and two different sets of longitudinal striations (*boxed regions*). Scale bar is 200 Å. **b** Fourier transform of the *upper boxed region* in (**a**), with the first row line at 18 Å (*arrow*). **c** Projecting along the [100] axis (*arrows*) causes alignment of the 18 Å [100] planes of the unit cell. *Small circles* represent α-helices within the coiled coil (*large grey circles*), the hexagonal unit cell with sides of 20.8 Å is shown in *orange*. **d** Fourier transform of the *lower boxed region* in (**a**), with the row line at 10.4 Å (*arrow*). In both (**b**) and (**d**) the first meridional reflection is at 41.8 Å. **e** Projecting along the [110] axis aligns the 10.4 Å [110] planes

striations. Images with both SAF and T4 present allowed the lateral striations to be determined as 41.8 ± 0.5 Å (S.D., n = 3).

A further level of order was present in cryoTEM images, and which had not been seen before, in the form of longitudinal striations that run parallel to the long axis of the fibre (Fig. 4.1a, b). It is important to note here that the images are formed from electron density of the protein and solvent only, so these striations are a direct visualisation of the superstructure of the fibre. Visibility of these striations in fibre projection images requires exquisite alignment of the coiled-coil building blocks within the fibre. Since the lateral striations were also visible they could be used to calibrate the width of the longitudinal striations as 18 ± 0.6 Å (S.D., n = 25).

Since it is known that the coiled coils are hexagonally packed within the SAF superstructure [2], this led to the conclusion that the high density regions in the new images were formed by projection of the [100] planes of the hexagonal array (Fig. 4.1b, c). The 18 Å striations indicate that the coiled coils are 20.8 Å apart (18 Å ÷ sin 60°). This value is larger than that found using WAX diffraction of 18.24 Å [2], a discrepancy probably caused by the partially dried state for which the WAX diffraction data was collected causing some contraction due to water loss, compared to the fully hydrated fibres imaged by cryoTEM.

With hexagonal packing, certain lattice planes are visible at specific projection angles through the fibres (Table 4.1, and refer to Fig. 3.2). The resolution of the EM is such that the [110] planes should also be visible. These would have a periodicity of half of the unit cell length. As predicted, some of the fibres also exhibited longitudinal striations with a much higher frequency (Fig. 4.1d, e). These were 10.4 Å apart, half of the hexagonal lattice distance (20.8 Å × cos 60°), and confirmed the hypothesis that these longitudinal striations are caused by the hexagonal packing. A square-packing regime, or any other array without planes separated by 120°, would not allow these ratios.

To ensure that the images were indeed of 3D fibres and not 2D sheets, electron cryo-tomography (cryoET) was performed on the SAFs (Fig. 4.2). The tomogram showed that the SAFs have approximately circular cross sections with uniform internal electron density; i.e., they are solid cylinders. This means that the [100] and [110] planes are visible through many layers of highly aligned coiled coils. This degree of order is remarkable in projection, since each SAF comprises some 30×10^6 peptides:

Table 4.1 Lattice plane spacing

Plane	Ratio	Spacing (Å)	Observed?
Unit cell	-	20.8	-
[100]	$\frac{1}{\sqrt{1}}$	18	✓
[110]	$\frac{1}{\sqrt{3}}$	10.4	✓
[200]	$\frac{1}{\sqrt{4}}$	9	✓
[210]	$\frac{1}{\sqrt{7}}$	6.8	×
[300]	$\frac{1}{\sqrt{9}}$	6	✓

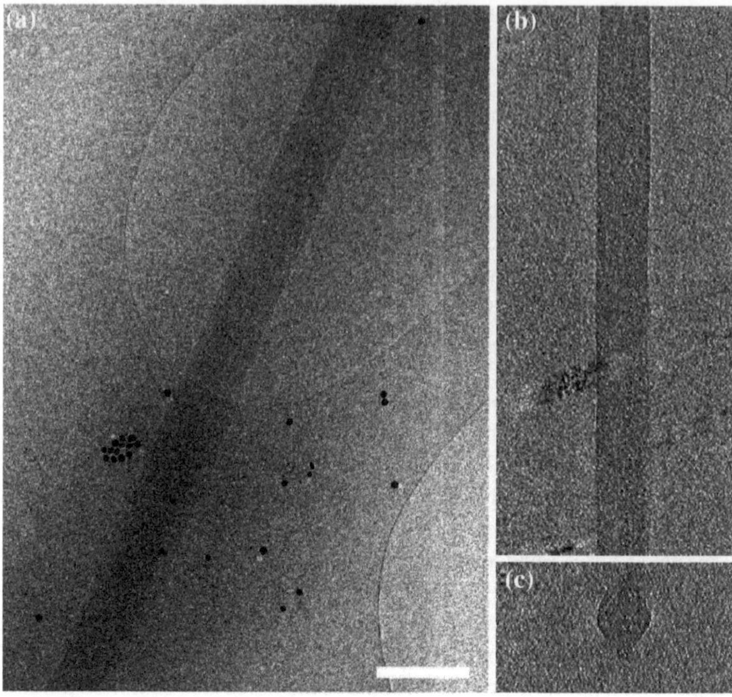

Fig. 4.2 A tomogram of frozen hydrated SAFs. **a** Image of a SAF used to collect a tomogram from ±45°. The lacey carbon and gold fiducial markers used to aid reconstruction are clearly visible. **b** Tomographic slice through the fibre in (**a**) showing the uniform density present in the interior of the SAF. The image is an average of 30 tomographic slices to aid contrast. **c** Fibre cross-section showing the approximately circular cross-section of the SAF. Scale bar is 200 nm for all images

Assuming the SAFs are perfect cylinders comprising coiled coils with six peptides per pitch and hexagonally packed with maximum efficiency $\left(\frac{\pi}{\sqrt{12}}\right)$, the number of peptides (N^o_{pep}) will be equal to:

$$N^o_{pep} = \frac{6\pi}{\sqrt{12}}\left(\frac{\pi r^2_{SAF} h}{\pi r^2_{cc} P}\right) \tag{4.1}$$

Where r_{SAF} is the average radius of the SAF (42 nm) and h is the average length of the SAF (42 μm) [5], and r_{cc} is the radius (1.04 nm) and P is the pitch (12.54 nm) of the composite coiled coils.

Although the lateral striations were observed ubiquitously along the SAFs, the longitudinal striations were apparent only at certain regions along the SAF (Fig. 4.1). The distance along the fibre long (z) axis between the regions of longitudinal striations was not a regular distance. This may imply either that there is a variable super-helical pitch or that the SAFs are distorted perpendicular to the plane of the image. Although cryoET showed that the SAFs have approximately circular cross sections it also showed that the cross section was randomly distorted. CryoET also showed that the SAFs were not perfectly straight and could therefore withstand some bending, which would occlude the longitudinal striations. These observations suggest that the SAFs resemble paracrystals more than either crystals or fibres; there is considerable short-range order but less long-range order. This would explain why the [110] planes are visible in the same view as the [100] planes: Even though the SAFs are straight, the coiled-coils are not perfectly aligned along the fibres, which is why the "unit cells" in Fig. 4.1c and e appear different. The hexagonal packing of the coiled coils (large circles) have rotated by 30° between Fig. 4.1c and e, which is due to a twist in the fibre itself and is not part of the short-range order of the unit cell. This essentially describes a system where numerous crystalline regions are present in different orientations along the fibre.

Tomography can be used to generate 3D electron density maps of protein complexes. Unfortunately, the magnification required to image the longitudinal striations (×50 k) generated an electron flux that destroyed the high-resolution data after only one image, preventing the use of high-resolution tomography in this study. Rotating the SAFs around the fibre long-axis should align different planes with the viewing axis of the EM. At the resolution and magnification used, there should be three distinct sets of planes visible (Table 4.1). As shown above, the [100] and [110] planes are visible (Fig. 4.1), as are the [200] and [300] planes in some micrographs. However, the [210] planes were not discovered. This may have been due to the information limit of the EM used. The Tecnai T20 is equipped with a LaB_6 crystal filament which, although much brighter than a tungsten filament, has a much lower achievable resolution than that possible with a FEG [6]. It may have also been due to the choice of defocus used during image acquisition, which would have precluded the observation of certain Bragg reflections, as will be discussed below. The [300] planes were only visible in some SAFs, and were more obvious due to their coincidence with the [100] planes, which were relatively straightforward to image. Since the striations were visible in projection through many layers of coiled coils even a small rotation or tilt of the [100] or [110] planes relative to the electron beam would have resulted in a complete loss of visibility: A rotation of 1.2° would be enough to obscure the 18 Å striations.

Since the SAFs are so exquisitely aligned, methods designed for the analysis of 2D crystals should be applicable to solve the structure of the 3D fibre.

4.2 An Electron Density Map of the SAF Ultrastructure

Computing the FT of the fibres gave a single crystallographic lattice with equatorial reflections out to 6 Å and meridional reflections out to at least 20.9 Å (Fig. 4.3). The meridional reflections describe features along the fibre long axis, whereas equatorial reflections describe the lateral hexagonal packing of the coiled coils. The first layer line of the FT was at 125.4 Å, corresponding to the pitch of the coiled coil. This

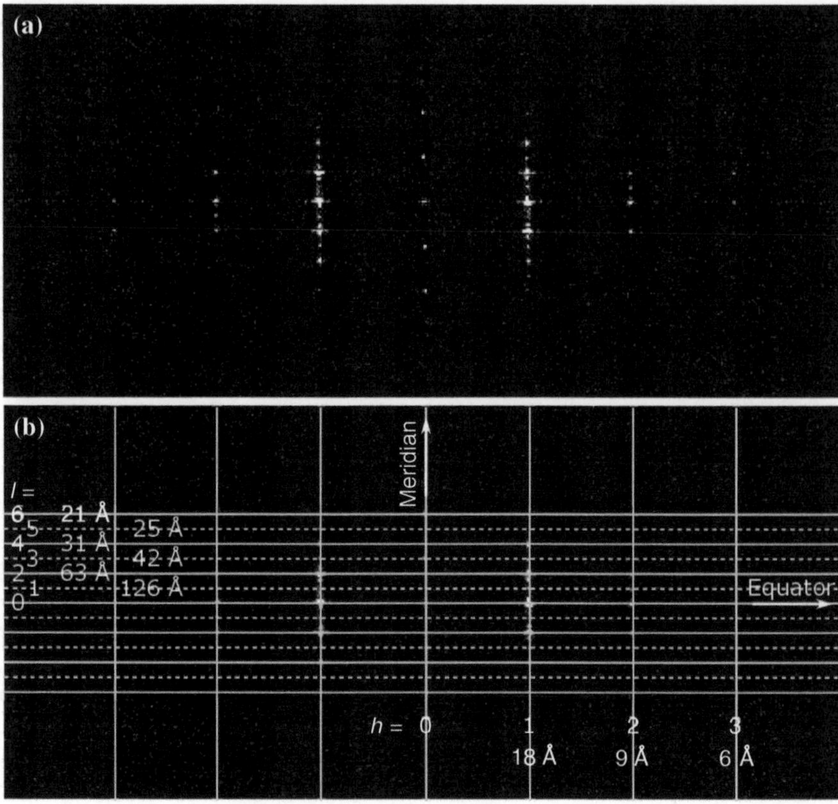

Fig. 4.3 FT of the SAF striations. **a** Background-subtracted FT of the SAF in Fig. 4.6 annotated (**b**) with *layer* and *row lines*. The *layer lines* (l) extend at least to the sixth which is at 20.9 Å, whereas *row lines* (h) extend out to the third at 6 Å

value fits with the pitch of coiled coils in nature, which varies from 80–200 Å, as judged from data mining coiled-coil structures found within the CC+ database [7] and analysed with the program Twister [8]. The third order meridional at 41.8 Å indicated that the SAFs are composed of a 3_1 helix, which means that the peptides repeat three times in one pitch of the coiled coil.

The fact that the peptides occur three times in a single pitch explains the hexagonal packing, since repeating motifs will occur every 120°. This allows repeating side-chain mediated inter-fibril interactions that cement the hexagonal packing, which could not be achieved with a non-integer, or indeed any other superhelix (Fig. 4.4). The longitudinal striations appear in lines of high density separated by stripes of low density (Fig. 4.1a). This phenomenon is caused by the pitch of the coiled coils; as the α-helices trace out the coiled coil they spread out so the electron density becomes more uniform across the [100] plane. Conversely, as the α-helices align with the [100]

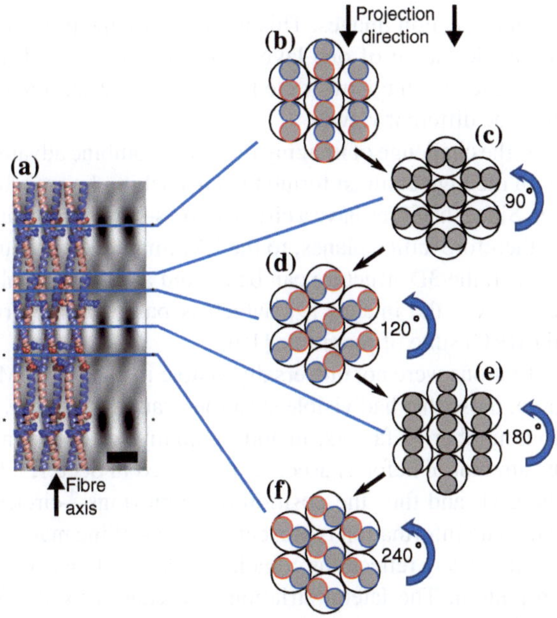

Fig. 4.4 Explanation of the longitudinal and lateral striations. **a** A Fourier-filtered image of a portion of SAF exhibiting lateral and longitudinal striations caused by alignment of the [100] planes. Protein density is *dark*. The cartoon overlay shows a basic coiled coil model, where α-helices are *coloured* as in Fig. 1.24. Arginine and aspartic acid side-chains are shown as spheres. Scale bar is 20 Å. **b** α-helices are shown as *filled circles* within the coiled coil (*large black circles*). Projecting through the SAF at the level of the *horizontal bar* shows basic and acidic residues (*blue and red circles*, respectively) coinciding with a lateral striation (*asterisk*) in the Fourier-filtered image, whereas the longitudinal striations are caused by projecting along the [100] axis. **c** After 31.35 Å along the SAF from (**a**) the α-helices have rotated 90° within the coiled coil. The longitudinal striations are no longer clear since the α-helices have spread out and the electron density is more uniform. **d** After 41.8 Å the α-helices have rotated 120° from (**b**) and there is another lateral striation coinciding with charged residues, although there are no longitudinal striations since the α-helices are not aligned with the imaging axis. **e** After half a pitch (63 Å) the α-helices are again aligned with the imaging axis and longitudinal striations result. Lateral striations occur once more after 83.6 Å (**f**) before the coiled coil has completed a full pitch

plane this leads to areas of high and low protein density, visible as the longitudinal striations (Fig. 4.4).

The SAFs are unique in the respect that they are helical filaments in-register forming a 3D array which can be treated as a 2D crystal. Helical filaments are ideal for 3D reconstruction since they yield the same structure from many views along one filament. However, since they are single filaments they present their own problems. Namely, thin filaments bend and have to be interpolated to a straight line, introducing artifacts. They also suffer from a low signal-to-noise ratio (SNR). As a consequence, although structures at 20 Å resolution are achieved regularly, only a handful have been solved to better than 8 Å. 3D arrays, however, are usually much more complex for 3D reconstruction from projection images. This is caused by the problem of separating out the crystallographic lattice of one layer from the next, which means that the number of lattice planes visible is restricted to a small subset, unless tomography can be used to provide different views.

Serendipitously, the challenge of imaging the SAFs combine advantages of helical reconstruction with the high contrast formed from stacking layers of coiled coils on top of one another. Since the fibres have a circular cross-section they present different orientations, and therefore lattice planes, to the EM imaging axis. Due to the coiled coils being in register, the 3D structure can be treated as a 2D crystal. The program 2dx was used to process the images [9], which is based on the original medical research council (MRC) suite of programs [10].

Longitudinal striations were not universally visible in all cryoTEM images of the SAFs. Furthermore, those that had visible striations varied in quality. This variation was attributed to a number of factors, including quality of fibres and thickness of ice, as well as the amount of defocus used. As described in Sect. 1.3.2, changing the defocus alters the CTF and thus the position of Thon rings—circles in reciprocal space between which no information is present. For crystalline materials the defocus must be set so that the Bragg reflections coincide with the Thon rings and contribute to the diffraction pattern. The lateral striations are caused by the 3_1 coiled coils comprising the SAFs and so are henceforth referred to as the [003] planes. The [100] lattice planes need to be visible under the same defocus setting as the [003] planes.

For the cryoTEM images presented in this Thesis the defocus was set to $-1.47\,\mu$m. This caused zeroes of the CTF to appear at 19.4, 13.6, 11.1, 9.6, 8.6, 7.8, 7.3, 6.8, 6.4, 6.1 and 5.8 Å (Fig. 4.5). This may be the reason why the [210] planes at 6.8 Å have not been imaged whilst the [300] planes at 6 Å have; they lie directly on a zero of the CTF.

The CTF must be corrected to extend the interpretation of the image beyond the point resolution. As previously mentioned, the point resolution is the resolution at which the image cannot be directly interpreted due to oscillations in the phases (black becomes white and *vice versa*). Correcting the CTF yielded images that had a lateral resolution of 6 Å (Fig. 4.6). At this resolution, electron density consistent with individual α-helices within dimeric coiled coils are clearly visible, as are the lateral striations, which was necessary for model fitting, discussed below. Due to the rapid oscillations of the CTF at high frequency, acquiring an image with diffraction spots past 9 Å along the equator proved to be difficult. A defocus value of $-1.47\,\mu$m was used, which allowed the visualisation of both the lateral and longitudinal striations. However, a defocus of $-0.5\,\mu$m should also prove to be a good value, allowing the pass of more high-resolution reflections and easing CTF determination and correction (Fig. 4.5). Unfortunately I had no opportunity to do this prior to completion of this Thesis.

After correcting the images for the CTF, the 2D electron density projection clearly showed density consistent with paired α-helices in the coiled-coil fibrils (Fig. 4.6), in agreement with the original design. Although the resolution of the image was not high enough to resolve the side chains of amino acids, the position of the lateral striations indicate which side chains were present; the lateral striations strongly suggested a

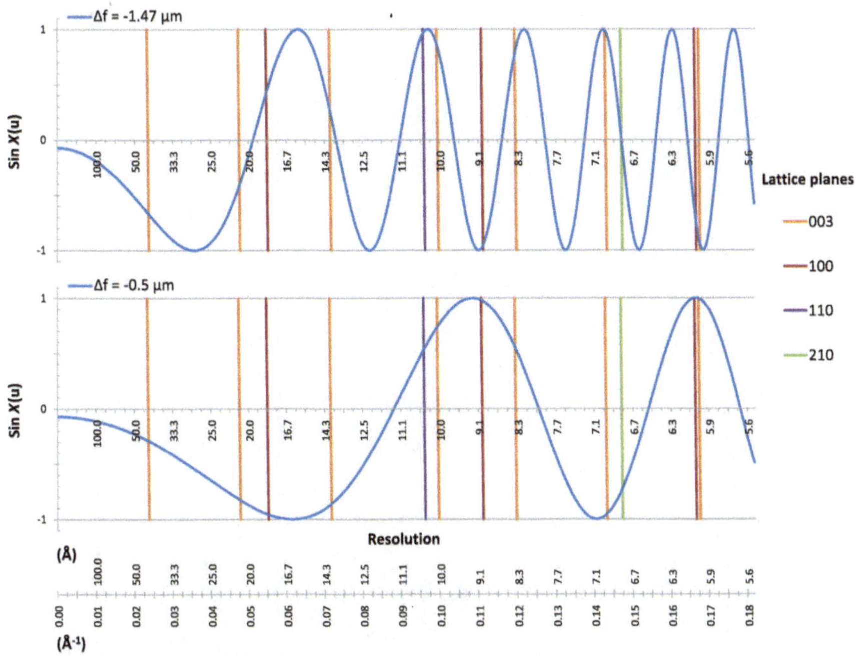

Fig. 4.5 Simulated CTF for defocus (Δf) $-1.47\,\mu$m and $-0.5\,\mu$m, shown with the position of the lattice planes of the SAFs. The *scale bar* shows resolution as a function of real (Å) and reciprocal ($Å^{-1}$) space. (Cs = 2 mm, E_0 = 200 kV, amplitude contrast = 7 %)

greater electron density between coiled coils. The sequences of the SAF peptides includes a group of aspartic acid (Asp, D) and arginine (Arg, R) residues that are opposite each other in the fully folded coiled coil (Table 4.2, underlined) and repeat three times per pitch, or once every 41.8 Å, and as such are ideal candidates for causing the lateral striations and mediating hexagonal packing by inter-coiled coil salt bridge formation.

Using cryoTEM images to compute the FT is generally more desirable than using electron diffraction since the phases of the reflections can be deduced as well as the intensity and direction. However, the CTF is a phenomenon of the objective lens, so although the phases can be determined, the intensity of the spots is modified by the CTF. Electron diffraction avoids this since the objective lens is not used to create contrast, it functions simply to focus the diffraction pattern to the imaging plane rather than the back focal plane. Combining these two modalities would increase the resolution of the final map since the intensity of the diffraction spots would be used as the amplitudes of the Bragg reflections, with the phases taken from the image. Forming a diffraction pattern requires a selected-area diffraction aperture, which unfortunately our Tecnai T20 does not have.

Another method used to increase the resolution of the final image is to merge images taken at different defocus values so that any spots that are absent due to the

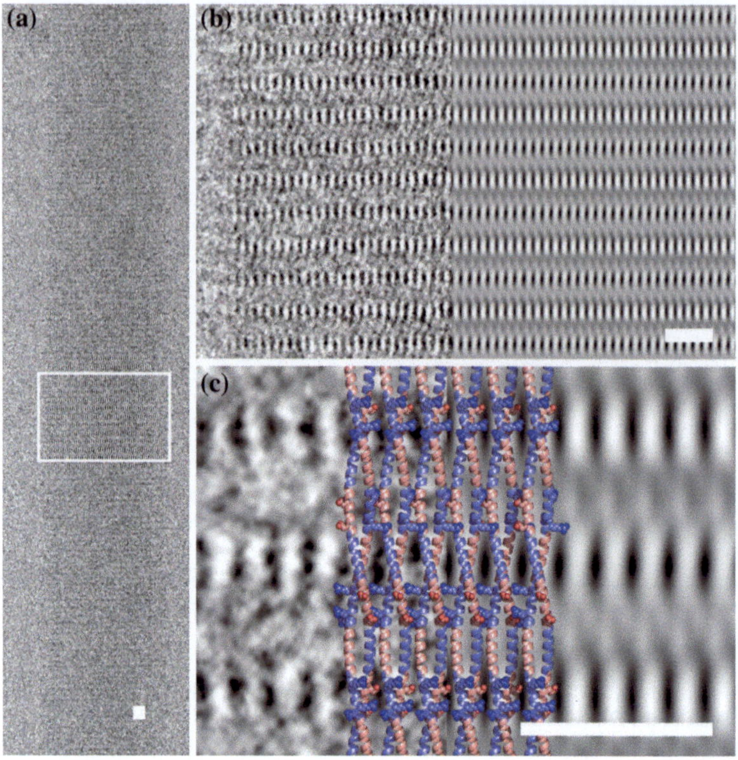

Fig. 4.6 Direct imaging of dimeric α-helical coiled coils within the SAF ultrastructure. **a** Region showing [100] planes to high resolution used for further analysis in this Thesis. **b** *Magnified region* showing the near-perfect lattice of high contrast striations with the Fourier-filtered CTF-corrected image (*right*) revealing the lateral striations spaced 41.8 Å apart. **c** CTF-corrected image (*right*), with a cartoon overlay of the postulated SAF structure. Heptads are coloured by overall charge (*blue and pink* are basic and acidic, respectively). *Light and dark* patterns indicate high and low electron density, respectively. The lateral striations are proposed to be due to the charged Arg and Asp residues (shown as *spheres*) at the *b* and *c* positions of the heptad repeat forming salt bridges between adjacent coiled coils. Scale bars are all 10 nm

Table 4.2 Sticky ended SAF peptide sequences.

Register	g abcdefg abcdefg abcdefg abcdefg abcdefg abcdef
SAF-p1	K IAALKQK IASLKQE IDALEYE NDALEQK IAALKQK IASLKQ
SAF-p2a	E IAALEYE IAALEQK IRRLKQK NARLKQE IAALEYE IAALEQ

The D and R residues residues that are opposite each other in the fully folded coiled coil are underlined. Black residues are the individual peptides, grey residues are shown to highlight the sticky ended assembly. See also Table 2.1 and Fig. 4.12

CTF (such as the [210] set above) are visible after the Thon rings have moved. This was not possible due to the beam damage to the fibres caused from multiple exposures. However, since the SAFs are presumed to be identical, it should be possible to merge images of different fibres to increase the SNR and increase the resolution.

The SAFs have longitudinal polarity; α-helices have an N- and a C- terminus and the coiled coils comprise parallel α-helices. This is compounded by the fact that the coiled coils are heterodimers, and so have radial as well as longitudinal polarity. As such, two mirror planes pass through each SAF image so the fibres were in one of four unknown orientations; up, down, mirror up, and mirror down. Orientating the SAF was therefore a prerequisite for merging. The problem is alleviated somewhat since the SAFs are composed of 3_1 coiled-coil helices the [100] planes should appear identical from three directions. This problem is by no means unique to the SAFs: Microtubules have both longitudinal and radial polarity and the task of discriminating one view from another causes much information to be discarded [11]. For microtubules the problem was surmounted by imaging the binding site for Taxol, which binds only to the β subunit [12]. However, in the case of the SAFs the heterodimeric peptides appear identical at the resolution of the EM, and information at much higher resolution is required to discern between the four populations than those of tubulin.

In an attempt to merge the data from the images, a subset of images were chosen with good quality values (*QVal*). *QVal* is a measure of quality based on SNR of the diffraction spots of the entire image [9]. The Fourier-filtered and CTF-corrected images were then processed in one of two ways. Firstly, each image was aligned by eye to the image with the highest *QVal*. This involved changing the phase origin of the image so that the lateral and longitudinal striations matched (i.e., centering the unit cell). These were then allowed to sample all four orientations to find the lowest phase residual. Briefly, this involved reducing the difference in phase of the diffraction spots to a minimum. A high-resolution cutoff of 9 Å was used so that noisy higher resolution data did not skew the outcome. A spot IQ (image quality, similar to *QVal* above but based on information local to the spot) cutoff was also applied which stopped spots with a low SNR affecting the data. This process was then repeated until the images were in the optimum orientation before the data were merged. The result was not convincing; the phases did not converge on a global minimum and so refining the data in this manner was abandoned. Secondly, a model of the SAFs generated from the non-merged data (described in detail below) was used as the refined model to which the maps were aligned. However, this approach also failed for the same reason as above.

The typical method for aligning and averaging images is to first generate a reference image using the best data. This is then used to align subsequent images which are then merged and used as the new reference. This procedure is iterated until convergence of the data [13]. However, since these data did not converge, merging was not continued and instead a model of the SAFs was generated and used to test the predictions made (Sect. 4.4).

4.3 X-ray Crystallography

Simultaneously to the cryoTEM studies, peptides were prepared for crystallisation and X-ray diffraction. This work was performed in collaboration with Drs. Nathan Zaccai, Marc Bruning and Andrew Thomson at the University of Bristol. Since the SAFs comprise sticky ended peptides that form fibres, not large crystals amenable to crystallographic analysis, the individual peptides were circularly permuted to form blunt-ended dimers (Blunt-p1-I and Blunt-p2a-I; Table 2.1, Fig. 4.7). An iodophenylalanine residue was substituted in place of a tyrosine, which was included to aid structure determination using the anomalous X-ray scattering from iodine at 1.7 Å. The propensity of the non-permuted iodinated peptide (SAF-p1-I, Table 2.1) to form fibres was confirmed by cryoTEM, which showed both lateral and longitudinal striations (Fig. 4.8a, b). Interestingly, after Fourier-filtering and CTF-correction the image of the fibres containing SAF-p1-I did not exhibit additional striations, which would be expected due to the high electron-density of iodine (Fig. 4.8c). The reason for this is that the iodophenylalanine occurs proximal to the aspartic acid and arginine residues, so any additional density will occur with the same pattern and spacing as the present lateral striations. This is therefore consistent with the model that the lateral striations are caused by the aspartic acid and arginine residues.

So far, the only permutation that has yielded crystals is the Blunt-p1-I:SAF-p2a heterodimer (Fig. 4.9a). The pitch and radius of the coiled coils present in the SAFs was determined to be 125.4 and 4.4 Å from cryoTEM images, whereas that of the crystal structure had an average pitch of 147.9 Å and radius of 4.9 Å (Fig. 4.9b). The smaller pitch for the coiled coils within the SAF with respect to the "free" coiled coils present in the crystal structure is likely due to the requirement of making hexagonal packing contacts. The fact that the pitch is smaller forces the heterodimers to repeat three times per pitch with an angular separation of 120°, ideal for hexagonal packing. The energy required to force this conformation could be offset by the lattice contacts made within the SAF array.

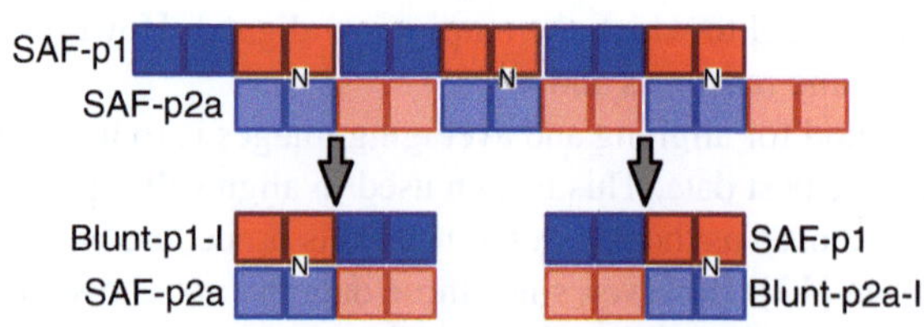

Fig. 4.7 Schematic representation of the circularly permuted peptides designed for crystallisation. *Upper model* is the standard sticky ended coiled coil configuration, *below* are the designed blunt ended blocks for crystallisation. *Blue and red squares* represent basic and acidic heptads, respectively, p1 is shown in *dark* with p2a in *light colours*. Breaks in the backbone are shown as gaps between *squares*. The position of the asparagine residue is denoted with N

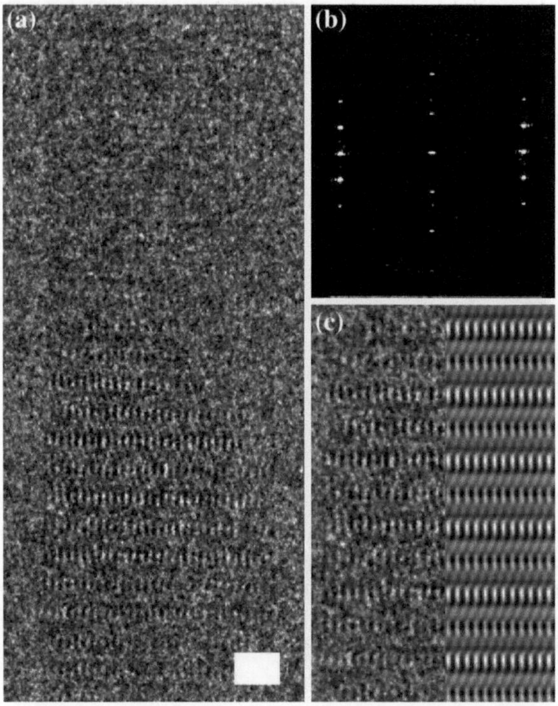

Fig. 4.8 Iodinated peptides are fibre-competent. **a** Fibre formed from SAF-p1-I and SAF-p2a (Table 2.1). **b** Fourier transform showing equatorial and meridional reflections identical to the fibre formed from non-iodinated peptides, confirming identical hexagonal packing. **c** CTF-corrected image of area of fibre in (**a**), showing the lateral striations every 41.8 Å. Scale bar is 10 nm for both (**a**) and (**c**)

4.4 Helical Reconstruction

Since the SAFs are helical filaments, the Fourier-filtered and CTF-corrected map was used to produce a 3D electron density map using helical reconstruction methods. As mentioned above, helical filaments are ideal candidates for 3D reconstruction since a single repeat presents the subunits from different, and known, angles.

Bessel orders of the layer lines were calculated based on the first maximum of the layer line out from the meridian, as well as prior knowledge of coiled-coil geometry (Fig. 4.10). The helical selection rule [14] is:

$$l = nK + mN \tag{4.2}$$

Where l is the layer line number, n is the Bessel order, K is the turns of the helix per axial repeat, m is an integer, and N is the number of units per axial repeat. So, for the SAF coiled coils Eq. (4.2) becomes:

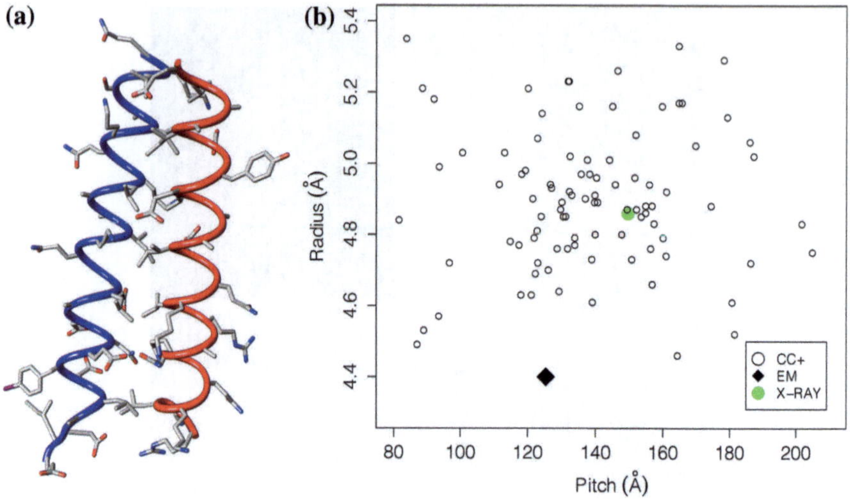

Fig. 4.9 Structural analysis of the standard and permuted SAF peptides. **a** X-ray crystal structure of the complex between Blunt-p1-I and SAF-p2a. The structure shows canonical coiled-coil interactions, including salt bridges and the hydrophobic core. **b** Plot of radius versus pitch for manually validated parallel dimeric coiled-coil structures deposited in the PDB. The parameters of the X-ray crystal structure of the Blunt-p1-I:SAF-p2a complex, and the model derived from cryoTEM are shown

$$l = n + 3m \tag{4.3}$$

However, due to the structural similarity between the heterodimeric α-helices, the coiled coil can be treated as a homodimer, equivalent to 2-fold cyclic symmetry about the z-axis. This restricts values of n to multiples of two [15]. The first layer line, corresponding to the pitch of the helix, corresponds to a two-start helix with Bessel order -2 (the minus sign denotes left-handed supercoiling), whereas the third is n = 0 and corresponds to the lateral striations ([003] planes) (Table 4.3). The resolution of the map was limited to 8 Å, this was necessary due to Bessel functions overlapping past 8 Å. For instance, on the second layer line in Fig. 4.10d there lies just outside the red ellipse a point of order n $= -4$.

The Fourier-filtered and CTF-corrected image shown in Fig. 4.6 was used for helical reconstruction. First, a single unit cell of the 2D map (Fig. 4.6) was used to generate a single filament of 10 helical pitches. This was floated and processed using the Brandeis Helical Package [16]. The resulting 3D map (Fig. 4.11a) clearly showed density consistent with a dimeric α-helical coiled coil, as well as lobes of electron density spaced 41.8 Å apart, into which an all-atom model could be placed.

A sticky ended model was produced based on data from the CTF-corrected map and the crystal structure. A program developed to generate continuous coiled coils called *makeccsc* (A gift from Dr. Gerald Offer, see also [17]) was used to produce a basic starting model of the SAF coiled coils. Since the coiled coils are 3_1 helices, there

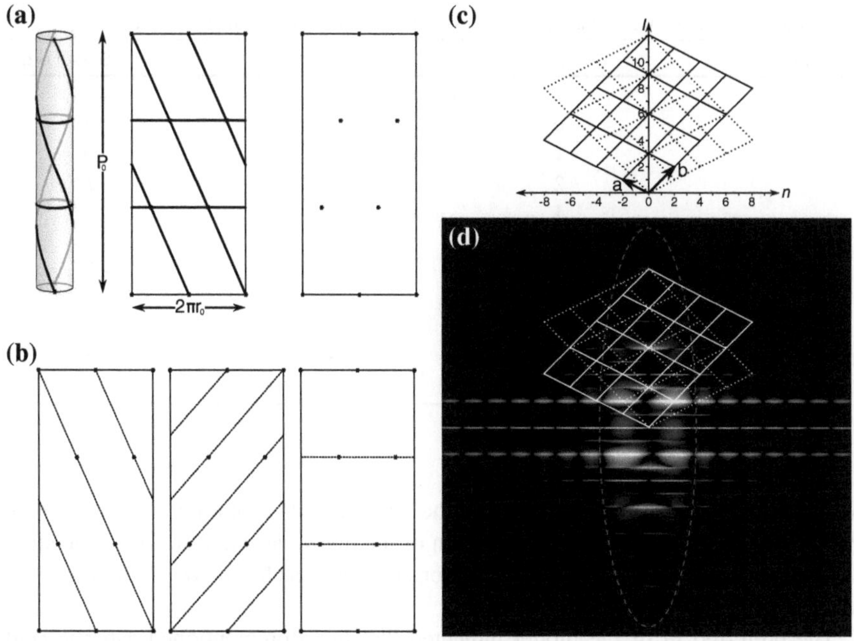

Fig. 4.10 Helical net and FT of the SAF coiled coils. **a** Helical net generation. The coiled coil is shown as a tube with the α-helices as lines coiling around it. The subunit translation distance, i.e., 41.8 Å, is marked with *circles* around the tube. Unfolding the tube reveals the 2D lattice, with the crossovers forming the lattice points. **b** The first three helices of order, from *left* to *right*, −2, 2, 0, with repeat distances of 125.4, 62.7, and 41.8 Å, respectively. **c** Reciprocal space lattice of the coiled coil, with basic vectors shown (**a, b**). The layerline number (l) and Bessel order (n) are shown. *Near and far* side lattices are shown as *solid and dashed lines*, respectively. **d** Lattice overlaid on the FT of the coiled coil. The zeros of the Bessel functions are clearly visible as *breaks* in the layer lines. The FT has been scaled by 25 % along the equator, so the *dotted red ellipse* traces out 8 Å^{-1}

are 84 residues in one pitch of the coiled coil and, due to the ~3.6 residues per turn in an α-helix, this means there are 23 turns of each α-helix. The radius of the center of the α-helices was measured from the images to be 4.4 Å from which, combined with the pitch, all parameters that govern coiled-coil geometry can be derived except the relative orientation of α-helices, which was set as 205°, as defined elsewhere [17]. The gross features were taken from the EM images, such as the pitch and width of the model, whereas the intra-coiled coil side-chain interactions were modelled based on the crystal structure data. Side chains thought to mediate inter-coiled coil interactions were left as extended rotamers. The breaks in the backbone formed from the sticky ended conformation of the peptides were added and the N- and C-termini rotated to remove steric clashes. The resulting model was docked as a rigid structure into the map using UCSF Chimera fitting routines [18]. A threshold value for the surface representation was set to enclose the volume corresponding to either 100 % (Fig. 4.11a) or 60 % (Fig. 4.11b) of the mass of the protein model (19,495 Da).

Table 4.3 Layer lines, Miller indices, and Bessel orders used for helical reconstruction

Layer line	*h*	*k*	Bessel order
1	1	0	−2
2	0	1	2
3	1	1	0
4	2	1	−2
5	1	2	2
6	2	2	0
7	3	2	−2
8	2	3	2
9	3	3	0
10	4	3	−2
12	4	4	0

A map was simulated from the atoms of the model at 8 Å resolution, which was then translated and rotated until the correlation of the maps was highest. This procedure was repeated using a number of starting orientations, each time returning the same solution.

The preferred fit of the model indicated that the aforementioned aspartic acid and arginine residues fill a large lobe of density, and is consistent with these being the cause of the lateral striations and the resultant hexagonal packing (Fig. 4.11c). In addition, there is also a proximal tyrosine residue, which may increase the local electron density, increasing the visibility of the lateral striations. Although the orientation of the polypeptide chains in the coiled-coil fibrils could not be discerned through this modelling, this did not affect the location of the charged and aromatic residues within the map.

Since the map was formed from the processed image of a single fibre the question of SNR remains. As before, processing the images in an attempt to produce averaged layer line data was not possible due to the difficulty in determining both axial and angular orientation. As described above, the resolution required to determine the orientation of the coiled coils is too high to merge the images. Therefore, I decided to build a model to test the predictions made so far and compare the electron density maps and FT patterns from the cryoTEM images to simulated data.

4.5 An All-Atom Model of the SAFs

Since the coiled coils within the SAFs are 3_1 helices, where the heterodimeric peptides repeat three times in one pitch of the coiled coil, this indicates that the hexagonal packing is a consequence of a specific structural moiety on the surface of the peptides repeating every 120°. As mentioned above, this moiety was identified as the charged aspartic acid and arginine residues which appear opposite each other once per

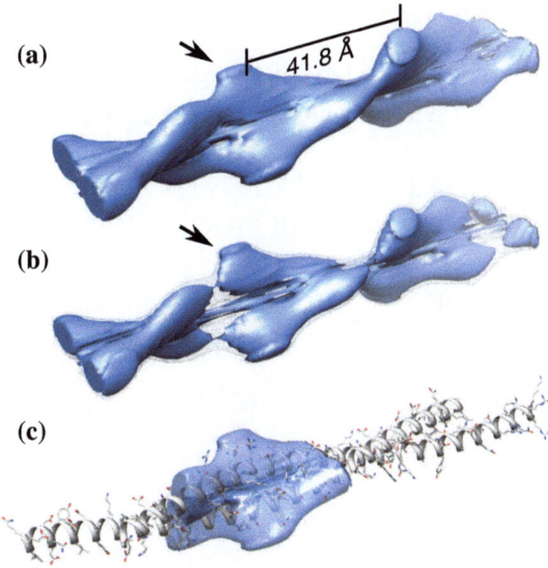

Fig. 4.11 3D electron density map of the composite coiled coils comprising the SAFs. **a** Electron density map with a surface threshold containing 100 % of the mass of the protein model. Lobes of density are spaced 41.8 Å apart. **b** Electron density surface with a threshold set to enclose 60 % of the mass of the protein model (100 % threshold is shown as a net). Electron density consistent with dimeric α-helices is visible. **c** The *arrows* indicate the lobe of electron density into which the D–R salt-bridges fit, as well as one of the tyrosine residues

peptide (Fig. 4.12). Figure 4.12 shows the deductive reasoning behind the proposed aspartic acid–arginine pairing; the hydrophobic core composed of isoleucine, leucine and asparagine is buried in the core of the coiled coil, whilst the alanine and serine residues are unlikely to form strong interactions and are present to primarily fill in the heptad positions and increase solvation. Similarly, the electrostatic interactions built into the e and g positions are paired with each other and therefore will not contribute to higher-order assembly. Although tyrosine and glutamine are able to form hydrogen bonds, their distribution reduces their competency to generate hexagonal packing in a 3_1 coiled coil; because of the repeat distance of tyrosine (21 Å) and glutamine (10.5 Å) the coiled coil will have rotated by either 90° or 180°, respectively. This leaves the three arginine and two aspartic acid residues. The repeat distance of these is ~42 Å, and as such they are ideal candidates for mediating hexagonal array formation since they repeat once every 120°.

This explains the visibility of the lateral striations in cryoTEM, where there is no stain present and contrast is generated by electron density of the protein and buffer only. Since the arginine and aspartic acid residues are likely to make electrostatic interactions, there will be strong electron density between adjacent coiled coils. One other such moiety that could have enough electron density to be seen in the EM are the tyrosine residues. However, these occur every 21 Å and therefore cannot be

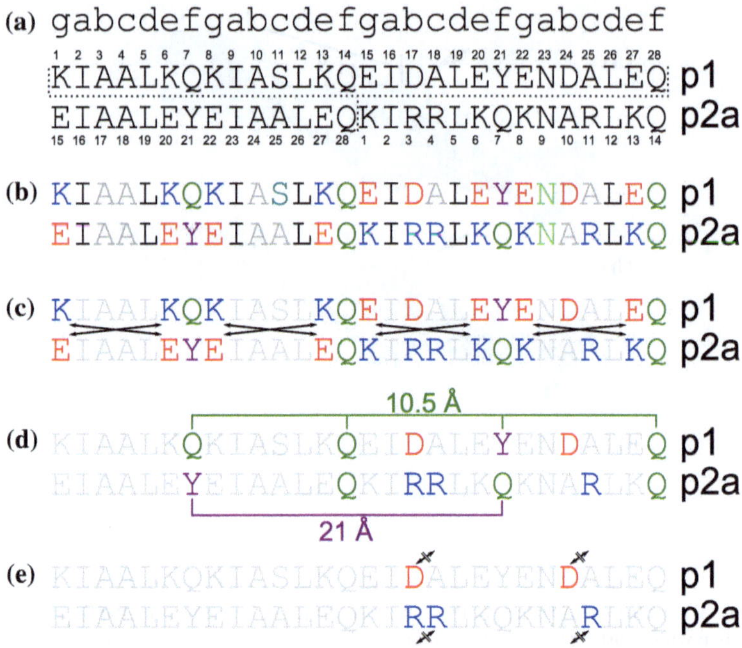

Fig. 4.12 Hexagonal packing and lateral striations related to the SAF peptide sequences. The heptad repeat is shown *above* the sequences. **a** SAF peptide sequences at the sticky-ended interface. Peptides are numbered and marqueed, so K_1 is the first residue in both peptides. **b** Residues coloured for identification only. *Blue* and *red* indicate basic and acidic residues, respectively. **c** *Grey* residues indicate satisfied interactions or residues that do not interact. *Arrows* show salt bridges formed between lysine and glutamic acid residues. **d** Glutamine (*green*) and tyrosine (*purple*) residue repeat distances. **e** Putative salt bridges forming between aspartic acid and arginine residues ($D_{17} \leftrightarrow R_4$ and $D_{24} \leftrightarrow R_{11}$) on adjacent coiled coils

causing the 41.8 Å striations. However, one of the tyrosine (Y) residues also occurs proximal to the aspartic acid–arginine salt-bridges, which together may generate enough electron density to be visible (Fig. 4.11c).

4.5.1 Molecular Dynamics

To visualise these putative interactions and probe their likelihood, a model of the sticky-ended coiled coil present in the SAF was produced with the calculated pitch and width. The side chains were modelled with those of the crystal structure (see Sect. 4.3) and the hexagonal array formed so that the aspartic acid–arginine salt bridges could form (Fig. 4.13a, b). To determine whether the all-atom model was stable this hexagonal conformation, the sticky ended model was subjected to molecular dynamics. This work was performed in collaboration with Dr. Richard Sessions at the

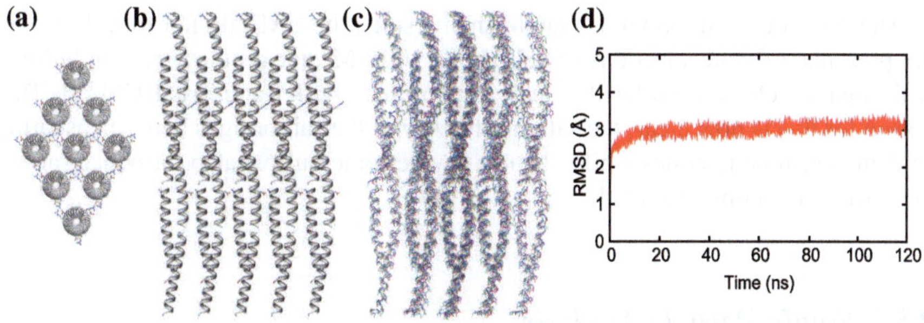

Fig. 4.13 Hexagonal array of coiled coils after 120 s of molecular dynamics simulation. The hexagonal array of coiled coils used for the MD simulations, shown from the *top* (**a**) and *bottom* (**b**) as a sticky ended unit cell, held together with Asp:Arg salt bridges. **c** View along the [100] axis of the array after 100 s of molecular dynamics. Colouring is by simulation condition: *Blue*, MOPS next to R_4; *Pink*, MOPS next to R_{11}; *Green*, Chloride ions only. **d** Root mean squared deviation of the peptide atoms with respect to the initial model. The graph indicates that the simulation is essentially equilibrated after 20 s and the size of the fluctuations are consistent with thermal motions rather than any major conformational change

University of Bristol. The model was placed into a periodic boundary box so that the sticky ends and electrostatic interactions persisted past the confines of the simulation box, as though the model was inside a SAF fibre. This was energy minimised and run on a simulator in the presence of MOPS [(3-(N-morpholino)propanesulfonic acid, the buffer in which the SAFs are grown] placed near either R_4 or R_{11}, to simulate the buffer that the SAFs are grown in, or with no MOPS and chloride ions present to make the overall charge of the system zero (Fig. 4.13c). After an equilibration period of 20 ns, the root mean squared deviation (RMSD) over all 25,569 backbone and side-chain atoms to the initial structure remained at 3 Å for the remaining 100 ns of the simulation (Fig. 4.13d), indicating a viable configuration for the coiled coils within assembled fibres. Interestingly, the pitch decreased to 123.5 Å and the radius increased to ~5 Å, possibly indicating the lack of a stabilising force in the model. Nonetheless, all three systems were fully stable with the [100] and [110] planes intact (Fig. 4.13c). Indeed the MOPS, instead of acting as a stabilising molecule as had been hypothesised, appeared trapped by the sieve-like conformation of the array, whereas the Cl^- ions were small enough to become randomly distributed.

There is an imbalance of charges present in the SAF peptide designs; p2a contains an extra +1 charge in the form of an arginine residue (R_3 or R_4, Fig. 4.12). Which one of these arginine residues binds to aspartic acid is important from a protein design perspective. Following the MD simulations, the distribution of D–R interactions was assessed. At the start of the simulation, the array simulated both $D_{17} \leftrightarrow R_4$ and $D_{24} \leftrightarrow R_{11}$ interactions. However, after 120 ns, during which time the RMSD had reached a plateau, the $D_{17} \leftrightarrow R_4$ electrostatics were far more prevalent, with 95 % still present, although 6 % of the possible $D_{17} \leftrightarrow R_3$ salt-bridges had formed. This was opposed to the $D_{24} \leftrightarrow R_{11}$, which were less common with only 20 % still formed, perhaps indicating a less important role for these interactions.

These data fit with previous unpublished results; the SAFs do not form fibres in the presence of concentrations of salt above 100 mM, nor do they form in buffers with separate charged molecules such as phosphate buffer and Tris·HCl [19]. The reason for this is now clear; the salt interferes with the salt bridges between aspartic acid and arginine residues whilst the non-zwitterionic buffers also introduce salt in the form of the conjugate acid.

4.5.2 Single-Particle Analysis

Since each micrograph represented a single and separate view of the projected density of a thick, fibrous assembly and not a thin 2D crystal, and due to the wide distribution of radial orientations that the cylindrical fibre can present to the electron beam, the images processed using 2dx could not be merged using classical electron crystallography techniques. Similarly, due to the width of the fibre, tilt angles could not be obtained. At this stage, single-particle analysis was also not a viable approach since (i) the fibres are not "single" particles; and (ii) the angular distribution of usable images was extremely narrow as most projection angles show no high-resolution detail (again, because of the width of the fibre), and because the aspect ratio of the fibres restricts meaningful views out of the horizontal imaging plane. Similarly, 2dx-processed images could not be used to refine the structure of a single coiled-coil fibril iteratively, because the projected density of multiple layers of highly aligned α-helices produced images that would not be visible with an individual coiled-coil fibril.

Instead, a hybrid approach was designed so that the images processed in 2dx were used as high-resolution "class-averages" (Fig. 4.14), which were then compared to the projected electron density of the all-atom model, which is identical to the model used for molecular dynamics but contains more coiled coils (Fig. 4.15a). Using EMAN2 [20], the model hexagonal array of coiled coils was used to simulate an electron density map filtered to 10 Å. The electron density of the model was projected over angular steps of 5° and the resulting projection images compared and aligned to the processed images output from 2dx (Fig. 4.15b). Interestingly, this model predicted that additional striations would appear in fibres when tilted out of the imaging plane, which correspond to projections that are only visible in 3D fibres and therefore could not be visible in images of individual coiled-coil fibrils. This prediction was verified by experimental images of fibres that did not have lateral striations (Fig. 4.14c, d), and were therefore not considered for initial model development. Hence, although lateral striations are visible over the full angular range around the fibre axis, they rapidly disappear as the fibre tilts away from the horizontal. Conversely, the longitudinal striations are visible over only a narrow in-plane rotation, but are visible over a larger out-of-plane tilt due to the alignment of α-helices.

Matching projection images and processed micrographs are shown in Fig. 4.15b, which illustrate the similarity between the two sets. Due to the fact that individual

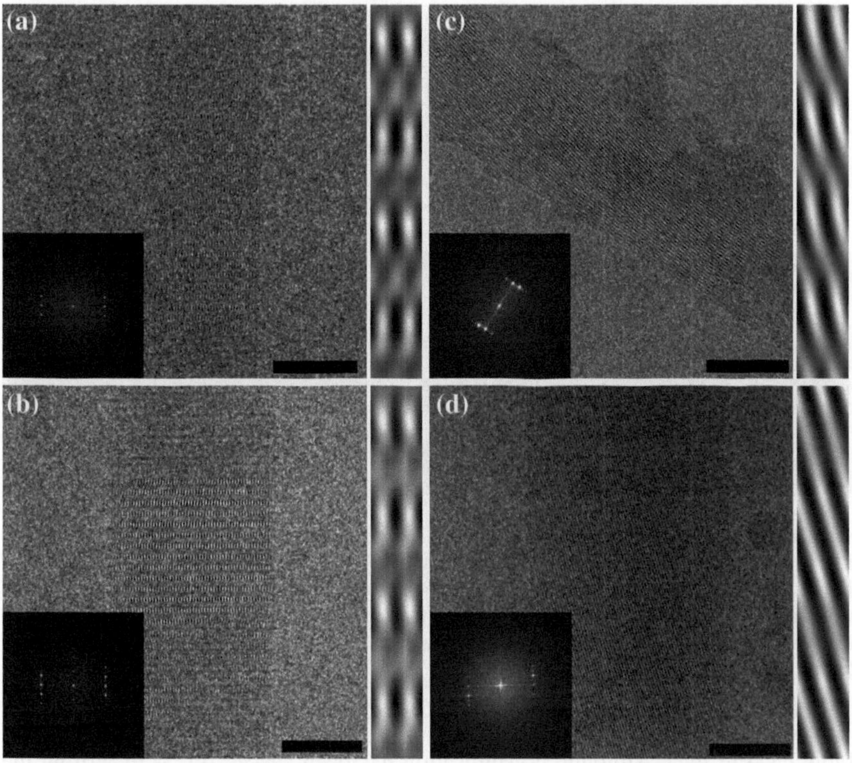

Fig. 4.14 A sub-selection of images processed by 2dx used for single-particle analysis. (**a**, **b**) with clear lateral and longitudinal striations. The *insets* show the computed FT, the first equatorial reflection is at 18 $Å^{-1}$. Processed 2 × 2 unit cells are shown to the *right* of each image. (**c**, **d**) Fibres with slanted striations, which match model data when the SAF is tilted out of the imaging plane (see Fig. 4.15b). All scale bars are 50 nm

α-helices are clearly visible in both projected and processed images, a conservative resolution estimate of 8 Å can be assigned to the data presented herein.

4.6 Future Work

Although the crystal structure shown in Fig. 4.9 provides most of the intra-coiled coil interactions, the conformations of the arginine residues are not well ordered in this structure (C_ζ temperature factor >63, *cf.* average C_α temperature factor = 42), probably due to the proximity of these residues to the edge of the unit cells and relatively high solvent-accessibility. To see whether the arginine residues behaved in an unusual way when buried, a separate circularly-permuted peptide was synthesised with the arginine residues in the center of the heterodimeric coiled coil (Blunt-p2a-I,

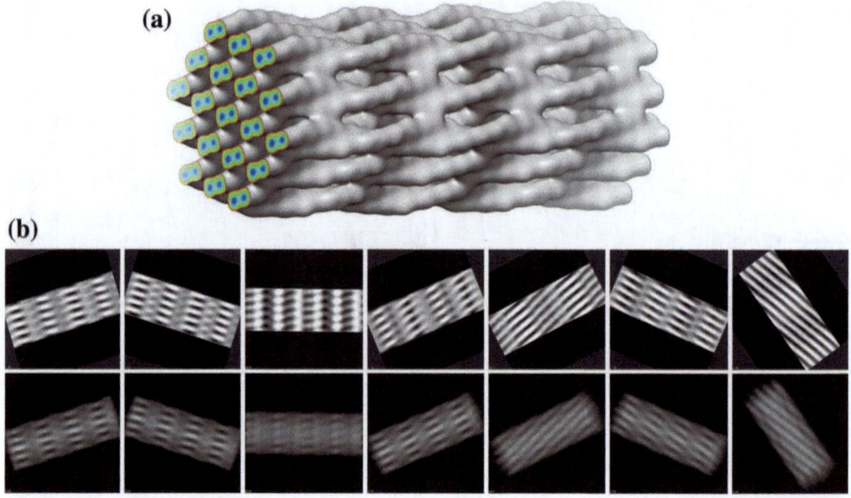

Fig. 4.15 Model of the SAF superstructure used to simulate cryoTEM images. **a** The model was filtered to 10 Å used to simulate projection images. The surface threshold is set to contain 100 % of the mass of the protein array. Protein occupies 50 % of the volume of the array and contains pores of approximately 100 $Å^2$. The *coloured plane* shows the density gradient and clearly illustrates the two α-helices comprising the coiled coils. **b** Projection images of the model (*bottom row*) and matching 2dx-processed images (*top row*)

Table 2.1, Fig. 4.7). Although crystallisation has been successful, as of writing the crystals have diffracted poorly and crystallisation optimisation is being performed.

Since the SAFs are also the first *sticky ended* fibre-forming system, the coordinates of the peptide termini were also sought. In has been proven that the termini are very close in space, since a mixture of amidated and acetylated peptides obviates fibrillogenesis [2]. Energy minimisation on simulated sticky ended peptides has also previously shown that the termini move only slightly [19]. In the simulation in this Thesis the termini were simply rotated to reduce steric clashes and allowed to reach equilibrium with the rest of the model. Due to this small, but significant, unknown in the structure, a second group of peptides was designed to crystallise in a sticky ended conformation, with two small peptides abridged by a longer peptide ($p1_3-N$, $p1_3-C$ and $p2a_6$; Table 2.1 Fig. 4.16). Crystallisation conditions tested so far have yielded only small fibrous crystals (Fig. 4.17) but requires optimisation to get diffraction quality crystals.

4.7 Discussion

Using a combination of cryoTEM, electron crystallography, helical reconstruction and single-particle analysis, the first high-resolution structure of a *de novo*-designed coiled coil has been solved, which has enabled an atomistic model to be built. Due

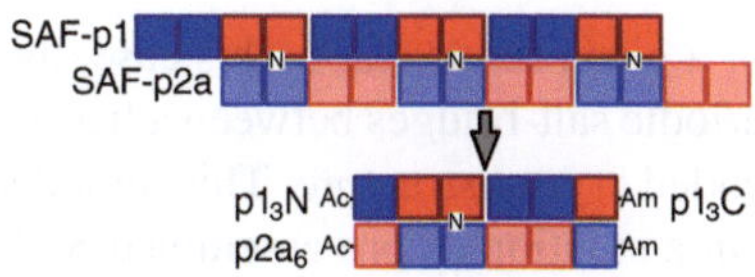

Fig. 4.16 Schematic representation of the sticky ended peptides designed for crystallisation. *Upper model* is the standard fibre-forming coiled coil configuration, with the sticky ended crystal configuration *below*. Colouring and notation is as Fig. 4.7, with *-Am* and *Ac-* representing the amidated and acetylated termini, respectively

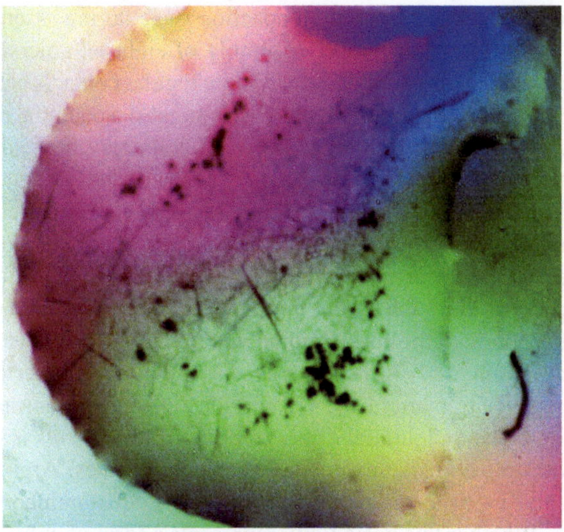

Fig. 4.17 Putative protein crystals for the trimeric sticky ended dimeric coiled coil. Crystals are grown using the hanging drop method and photographed under a polarising filter

to the unique structure of the SAF system, neither classical electron crystallographic techniques nor single-particle analysis could be used to analyze the micrographs. Instead, an approach was developed that combined analysis of individual micrographs using 2D crystallography, prior to comparing the images to projections from a 3D model, similar to single-particle analysis. With this information and an X-ray crystal structure of a related peptide complex, an all-atom model for the SAFs was generated and verified computationally. The model fits with all experimental data established thus far but, most importantly, adds to our knowledge of superstructure formation. This model is entirely stable to prolonged molecular dynamics simulation, confirming its structural viability. The pitch and width of the coiled coils has been determined for the first time in fully formed hydrated fibres. Furthermore, the interactions between adjacent coiled coils have been deduced to be salt bridges between a pair of proximal arginine and aspartic acid residues. These data are interrelated; the pitch and radius of the coiled coils within the SAF were smaller than those found in the crystal structure. The difference between the two structures means that pitch and radius are flexible for a given sequence. In this case, the pitch has decreased to allow the formation of periodic salt-bridges between adjacent coiled coils, which are not present in the blunt-ended crystal structure. This remodelling of structure possibly reflects the situation in a solution of pre-assembled SAF-peptide heterodimers, through to the assembled, paracrystalline SAFs [21].

This mechanism may also be applicable to the structure of intermediate filaments (IFs). Although it is known that IFs are predominantly coiled coil in structure [22], their structure within filaments has proven difficult to solve, partly due to structural heterogeneity and the length of each molecule prohibiting crystallisation. However, it is reported that IFs undergo a radial compaction stage during assembly [23]. This may reflect the transition from a loose assembly of eight tetramers to a more formal assembly, as the filaments form stronger, more regular contacts between adjacent coiled-coil fibrils. Results presented in this Thesis suggest a potential packing regime within the mature fibril, whereby adjacent coiled coils may be held together with salt-bridges that repeat with the same frequency as the coiled-coil pitch. IFs contain stutters and linker regions, which increase flexibility between the coiled coil domains. They are also highly charged, with a preponderance of acidic residues [22]. These two structural factors make the remodelling of coiled coils possible during polymerisation and compaction, although harder to predict since there is not a constant pitch. A recent crystal structure of the entire human vimentin coil 1b, shown in Fig. 4.18, is the first *bona fide* tetrameric IF structure solved (a previous tetrameric structure was found to be a crystallisation artefact [24]), and illustrates the importance of charge–charge interactions. The antiparallel dimer of dimeric coiled coils is predominantly held together with salt-bridges that occur at half-pitch values. Arginine is the sole basic residue, whilst aspartic and glutamic acid donate equal numbers of negatively charged residues. Whether this periodicity represents a form of cubic packing in the fully assembled IF remains to be seen; unfortunately, there is no publication that accompanies this structure as of writing.

These data increase the potential for dialing-in fibre thickness into nascent design principles. It also allows identification of regions within the primary sequence that are amenable to rational functionalisation. For instance, there are gaps between the coiled coils presumably filled with solvent which could be modified to form coordination sites for inorganic ions or catalysts, which could then act as molecular sieves, traps,

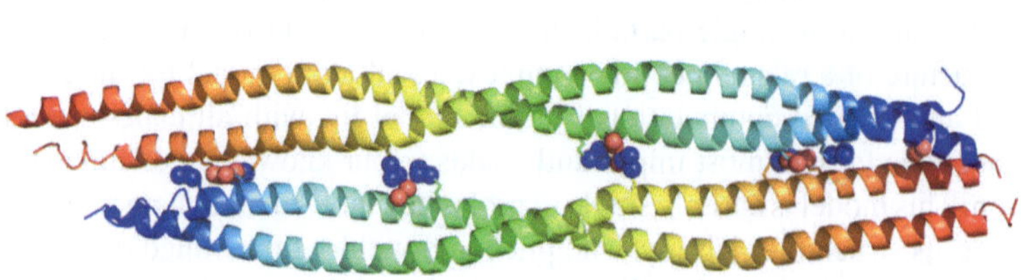

Fig. 4.18 Crystal structure of a tetrameric portion of human vimentin (PDB code 3OL1). The cartoon representation of the peptide backbone is shown *coloured* from *blue* N-termini to *red* C-termini. Arginine, aspartic acid and glutamic acid are shown as *sticks*, with the interacting N and O atoms shown as *spheres*. The interactions are predicted purely on proximity and orientation

or self-assembling solid-phase catalysts. The SAFs are a robust fibre-forming system with lateral striations reliably visible by both negative stain and under frozen hydrated conditions at a wide range of defocus, which also lends itself to use as a general purpose calibrant for high resolution TEM and cryoTEM.

Structures of previous *de novo* designed fibrous biomaterials have been solved by X-ray fibre diffraction and computer modelling combined with low resolution (∼30 Å) negative stain TEM data [25–28]. The structure presented herein is the first *de novo* designed fibre determined with atomistic detail, with specific interactions identified and computationally verified. These new structural insights, and the recently determined assembly pathway [21], mean that the SAFs represent the best understood designed, peptide-based, fibre-forming system presented to date.

References

1. T.H. Sharp, M. Bruning, J. Mantell, R.B. Sessions, A.R. Thomson, N.R. Zaccai, R.L. Brady, P. Verkade, D.N. Woolfson, Cryo-transmission electron microscopy structure of a gigadalton peptide fiber of de novo design. PNAS **109**(33), 13266–13271 (2012)
2. D. Papapostolou, A.M. Smith, E.D.T. Atkins, S.J. Oliver, M.G. Ryadnov, L.C. Serpell, D.N. Woolfson, Engineering nanoscale order into a designed protein fiber. PNAS **104**(26), 10853–10858 (2007)
3. N.G. Wrigley, The lattice spacing of crystalline catalase as an internal standard of length in electron microscopy. J. Ultrastruct. Res. **24**(5), 454–464 (1968)
4. M.F. Moody, Application of optical diffraction to helical structures in the bacteriophage tail. Philos. Trans. R. Soc. B Biol. Sci. **261**(837), 181–195 (1971)
5. S.C. Holmstro, P.J.S. King, M.G. Ryadnov, M.F. Butler, S. Mann, D.N. Woolfson, Templating silica nanostructures on rationally designed self-assembled peptide fibers. Langmuir **24**(20), 11778–11783 (2008)
6. A.F. De Jong, D. Van Dyck, Ultimate resolution and information in electron microscopy II. The information limit of transmission electron microscopes. Ultramicroscopy **49**(1–4), 66–80 (1993)
7. O.D. Testa, E. Moutevelis, D.N. Woolfson, CC+: a relational database of coiled-coil structures. Nucleic Acids Res. 37(Database issue), D315–322 (2009)
8. S.V. Strelkov, P. Burkhard, Analysis of alpha-helical coiled coils with the program TWISTER reveals a structural mechanism for stutter compensation. J. Struct. Biol. **137**(1–2), 54–64 (2002)
9. B. Gipson, X. Zeng, Z. Zhang, H. Stahlberg, 2dx-User-friendly image processing for 2D crystals. J. Struct. Biol. **157**(1), 64–72 (2007)
10. R.A. Crowther, R. Henderson, J.M. Smith, MRC image processing programs. J. Struct. Biol. **116**(1), 9–16 (1996)
11. T. Fujii, A.H. Iwane, T. Yanagida, K. Namba, Direct visualization of secondary structures of F-actin by electron cryomicroscopy. Nature **467**(7316), 724–728 (2010)
12. E. Nogales, S.G. Wolf, I.A. Khan, R.F. Ludueña, K.H. Downing, Structure of tubulin at 6.5 A and location of the taxol-binding site. Nature **375**(6530), 424–427 (1995)
13. E.V. Orlova, H.R. Saibil, Structural analysis of macromolecular assemblies by electron microscopy. Chem. Rev. **111**(12), 7710–7748 (2011)
14. M.F. Moody, *Image analysis of electron micrographs* (Academic Press, Massachusetts, 1990), 145–287
15. M. Stewart, Computer image processing of electron micrographs of biological structures with helical symmetry. J. Electron Microsc. Tech. **9**(4), 325–358 (1988)

16. C.H. Owen, D.G. Morgan, D.J. DeRosier, Image analysis of helical objects: the Brandeis Helical Package. J. Struct. Biol. **116**(1) 167–175 (1996)
17. G. Offer, R. Sessions, Computer modelling of the alpha-helical coiled coil: packing of side-chains in the inner core. J. Mol. Biol. **249**(5), 967–987 (1995)
18. T. Goddard, C. Huang, T. Ferrin, Visualizing density maps with UCSF chimera. J. Struct. Biol. **157**(1), 281–287 (2007)
19. A.M. Smith, *Rational Redesign and Characterisation of a Peptide-based Fibre*, Ph.D. thesis, University of Sussex, 2003
20. G. Tang, L. Peng, P. Baldwin, D. Mann, W. Jiang, I. Rees, S. Ludtke, EMAN2: an extensible image processing suite for electron microscopy. J. Struct. Biol. **157**(1), 38–46 (2007)
21. E.H. Bromley, K.J. Channon, P.J. King, Z.N. Mahmoud, E.F.Banwell, M.F. Butler, M.P. Crump, T.R. Dafforn, M.R. Hicks,J.D. Hirst, A. Rodger, D.N. Woolfson, Assembly pathway of a designed alpha-helical protein fiber. Biophys. J. **98**(8), 1668–1676 (2010)
22. H. Herrmann, U. Aebi, Intermediate filaments: molecular structure, assembly mechanism, and integration into functionally distinct intracellular scaffolds. Annu. Rev. Biochem. **73**(1), 749–789 (2004)
23. H. Herrmann, M. Häner, M. Brettel, N.-O. Ku, U. Aebi, Characterization of distinct early assembly units of different intermediate filament proteins. J. Mol. Biol. **286**(5), 1403–1420 (1999)
24. S. Nicolet, H. Herrmann, U. Aebi, S.V. Strelkov, Atomic structure of vimentin coil 2. J. Struct. Biol. **170**(2), 369–376 (2010)
25. M. Zhou, A.M. Smith, A.K. Das, N.W. Hodson, R.F. Collins, R.V. Ulijn, J.E. Gough, Self-assembled peptide-based hydrogels as scaffolds for anchorage-dependent cells. Biomaterials **30**(13), 2523–2530 (2009)
26. R.P. Nagarkar, R.A. Hule, D.J. Pochan, J.P. Schneider, De novo design of strand-swapped beta-hairpin hydrogels. J. Am. Chem. Society **130**(13), 4466–4474 (2008)
27. J.D. Hartgerink, E. Beniash, S.I. Stupp, Self-assembly and mineralization of peptide-amphiphile nanofibers. Science **294**(5547), 1684–1688 (2001)
28. M. Bruning, L. Kreplak, S. Leopoldseder, S.A. Muller, P. Ringler, L. Duchesne, D.G. Fernig, A. Engel, Z. Ucurum-Fotiadis, O. Mayans, Bipartite design of a self-fibrillating protein copolymer with nanopatterned peptide display capabilities. Nano Lett. **10**(11), 4533–4537 (2010)

Chapter 5
Developing Novel Probes for Correlative Light Electron Microscopy

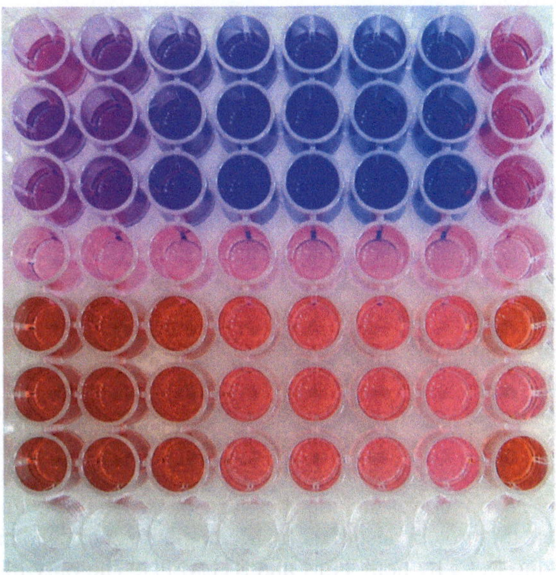

Interest in correlative light electron microscopy (CLEM) started in the 1950s but is now experiencing a resurgence, primarily due to the discovery of GFP and increased access to instruments required to achieve the technically demanding techniques.

A description of the generation of a fluorescent and electron dense protein by concatenation of metallothionein (MT), which was cloned adjacent to EGFP, is presented in Sect. 5.1. To elucidate the metal-binding capabilities of EGFP-MT2, purified protein was subjected to matrix-assisted laser desorption/ionisation (MALDI) mass spectrometry. The size of the protein with and without metal incubation was also assessed by dynamic light scattering. Next, imaging *in vitro* was achieved by localising the protein by encapsulation within liposomes or by protein-mediated liposome membrane tubulation, described in Sect. 5.2. Imaging the protein *in vivo* was attempted by localisation with various protein tags, immunolabelling whole cells

T. H. Sharp, *Biomolecular Imaging at High Spatial and Temporal Resolution In Vitro and In Vivo*, Springer Theses, DOI: 10.1007/978-3-319-02159-1_5,

using the Tokuyasu method, and autometallography to enhance the signal of any protein-mediated nanoparticles (Sect. 5.3). This chapter concludes with a description of a system that is currently being designed, along with ideas for future work to address some of the drawbacks of developing a general-purpose probe for CLEM (Sect. 5.5).

5.1 *In Vitro* Characterisation of EGFP-MT2

Enhanced green fluorescent protein (EGFP) is autofluorescent and genetically encoded. A similar protein for use in the TEM would be advantageous for high-resolution structural studies. As described in Sect. 1.5.4, metallothionein (MT) may have such properties.

5.1.1 Cloning and Purification of EGFP-MT2

As shown by Mercogliano and DeRosier, concatenation of MT results in cooperative binding of gold and higher stoichiometric ratios than found in non-concatenated MT [1]. Although MT is only 6 kDa in mass, after fusion to GFP the whole construct is >30 kDa. For this reason, the smallest number of repeats of MT as possible was desirable and so a tandem dimer of MT (MT2) was chosen as the fusion tag.

Pure protein was required for biophysical characterisation. Due to the size of the proteins, varying from 30–110 kDa, and the amount required, bacterial expression was essential. A His_6 tag was added to fusion proteins to enable one-step purification (see Sect. 2.3). Two MT genes were subcloned adjacent to one another within the original vector before subcloning the whole MT2 domain after the gene encoding His-tagged EGFP. No linker was included between the two MT domains to recreate as closely as possible the protein described by Mercogliano et al. who have previously reported visualising MT2 by both TEM and STEM [1]. The only mutation was an alanine to valine substitution at the interface between the two MT domains, caused by the sequence of the restriction endonuclease sites in the DNA.

Although purification of His-tagged EGFP was straightforward with high yields, purifying EGFP-MT2 proved to be much more difficult. This was probably due to the high abundance of cysteine residues, which may oxidise to form disulphide bonds between molecules, prohibiting purification. To enable purification, the bacteria required a very low concentration of IPTG combined with overnight induction at a temperature between 18–22 °C for soluble expression to occur. The presence of 0.3 mM $ZnCl_2$ was essential to keep the product soluble, presumably enhancing folding of the expressed protein by binding to MT2 and inhibiting disulphide bridge formation. Without any one of these three factors, EGFP-MT2 was sequestered into inclusion bodies at the poles of each bacterium and yielded very low levels of protein

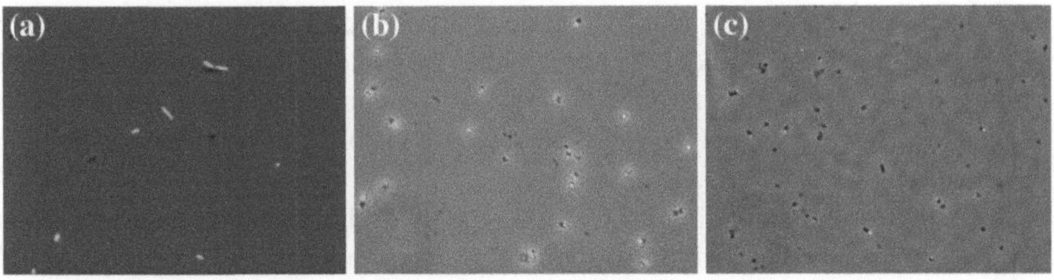

Fig. 5.1 Expression of EGFP and EGFP-MT2 in *E. coli*. EGFP with 1 mM IPTG (**a**), GFP-MT2 with 1 mM IPTG (**b**) or 0.05 mM IPTG (**c**)

(Fig. 5.1). Even with these factors inclusion bodies were still visible although relatively good yields of pure protein were still possible.

The presence of 20 mM tris(2-carboxyethyl)phosphine (TCEP), a phosphine based reducing agent, was essential during lysis of bacterial cells. β-mercaptoethanol (βME) was also tried but led to insoluble aggregates following lysis, possibly by forming cross-links between proteins *via* the high concentration of cysteine residues. The presence of a reducing agent following lysis is undesirable due to potential interactions with heavy metal compounds, and was found to be unnecessary.

Purification was achieved by using immobilised metal affinity chromatography. EGFP and EGFP-MT2 expressed from pET-45b(+) are tagged with an N-terminal His_6 motif. His_6 has high affinity for Ni^{2+} ions, which are chelated by nitrilotriacetic acid (NTA) bound to agarose beads. NTA is a tetradentate chelating moiety that coordinates four of the six ligand binding sites of Ni^{2+}; the other two sites bind with high affinity to the His_6 motif. Ni-NTA-agarose results in a high level of purification of His-tagged EGFP and EGFP-MT2 (Fig. 5.2). The predicted weights of His_6-EGFP and His_6-EGFP-MT2 are 30.7 and 42.0 kDa, respectively, and these can be clearly distinguished in Fig. 5.2.

Following purification, protein samples were desalted and the buffer exchanged for 25 mM Tris·HCl at pH 8 by size-exclusion spin columns. This was performed to remove traces of imidazole from purification. Various buffers were tested with

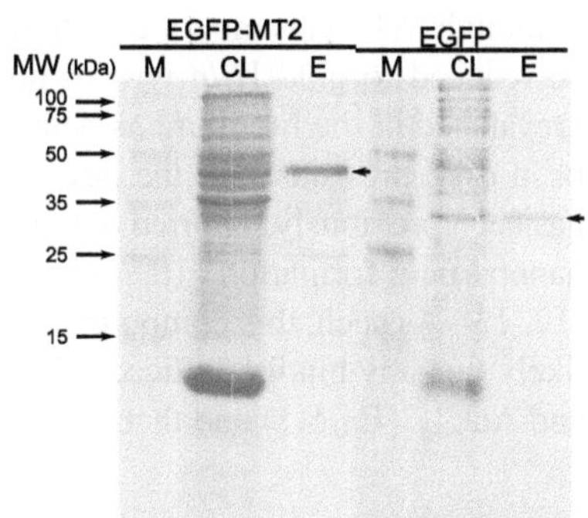

Fig. 5.2 Purification of His_6-tagged EGFP-MT2 and EGFP. *Arrows* indicate purified EGFP-MT2 and EGFP at 42 and 30.7 kDa, respectively. *M* Marker; *CL* Cleared lysate; *E* Eluate

both protein and the heavy metal compounds described below. Tris·HCl was the only buffer found that did not either precipitate the protein, or precipitate the metal compounds. A pH below 8 also resulted in the formation of bright-green insoluble precipitates. However, at pH 8 no precipitates were visible for extended time periods (>3 months).

Since multiprotein labelling for CLEM is highly desirable, larger concatenations of MT were also synthesised. It was thought that these higher concatenations would form larger nanoparticles visible and differentiable in the TEM. However, MT4, 6, 8, and 10 could not be purified, even in the presence of 20 mM TCEP, a reducing agent, and 8 M urea, a denaturing agent.

This strongly implies that the high concentration of cysteine residues forms disulphide bridges, leading to insoluble aggregates that render the protein intractable to purification.

5.1.2 Heavy Metal Compounds

To get the greatest electron density from the MT2 domain means binding to the heaviest metals possible. Complexes with various metal salts were attempted based on their solubility in aqueous buffers and previously reported interaction with MT. MT was originally classified as a cadmium and zinc binding protein [2]. For this reason cadmium chloride ($CdCl_2$) was used despite the high cytotoxicity of cadmium. $CdCl_2$ was also used by Nishino et al. and Fukunaga et al. [3, 4]. Platinum binds to MT as Pt(II) [5] and can displace cadmium [6]. Due to its good water solubility, K_2PtCl_4 was tested. Bismuth compounds bind to MT with the highest affinity of any metal tested [7]. However, bismuth compounds have low solubility in water and buffers suitable for cell culture, which precluded their use in this study. Gold is the preferred metal due to its relatively low cytotoxicity, high solubility of gold salts, and the fact that gold nanoparticles are already commonly used for immunolabelling for TEM. Both a gold salt, $KAuCl_4$, and a gold compound, Aurothiomalate (AuTM), were tested in this study. Because the aim was to develop a general probe for CLEM, $KAuCl_4$ was chosen due to the low cost of the salt, as well as its good water solubility. Although MT binds mono- and di-valent ions, trivalent gold is more soluble and can be reduced to gold(I) *in vivo* to bind MT [8]. AuTM was also chosen due to its prevalence in the literature as being an ideal reagent for MT [9] and the fact it has been used previously in the development of probes for TEM [1]. Although AuCl has been previously reported in the literature as a potential reagent for MT-mediated nanoparticle formation [10, 11], AuCl is extremely unstable in aqueous solutions [12, 13]. As such, this compound was briefly tested for solubility and discarded. It is likely that any binding efficacy shown by AuCl is due to disproportionation to Au^0 and $AuCl_4^-$ (Eq. 5.1) and therefore using $KAuCl_4$ instead should obviate its use.

$$4AuCl \rightleftharpoons 3Au^0 + AuCl_4^- \tag{5.1}$$

5.1.3 Mass Spectrometry

Matrix-assisted laser desorption/ionisation mass spectrometry (MALDI MS) was used to elucidate the mass-to-charge ratio of EGFP-MT2 before and after incubation with heavy-metal salts, and therefore calculate the number of ions bound to the protein.

The signal-to-noise ratio (SNR) varied widely depending on which matrix and buffer was used, whether either of the reducing agents TCEP or βME were present, as well as the concentration of protein and metal. Tris·HCl buffer (25 mM) at pH 8 was optimal, combined with 2,5-dihydroxybenzoic acid (DHB) with 0.1 % formic acid as a matrix. Even after optimisation, samples incubated with K_2PtCl_4 were not amenable to MALDI MS, and samples incubated with $KAuCl_4$ gave a very low SNR.

EGFP-MT2 has a predicted molecular mass of 42.012 kDa. However, when subjected to MALDI MS gave a peak at 42.309 kDa (pink line, Fig. 5.3), an increase of 297 Da. During expression in *E. coli*, $ZnCl_2$ was added to the growth media, which aided purification, possibly by improving the folding of the MT2 domain. The increase in mass may have reflected some $ZnCl_2$ bound to the protein, and indicated ($297 \div 65.39 =$) 4.5 bound Zn^{2+} ions. Similarly, the nickel column used for purification may leach some Ni^{2+} ions into the solution, and this peak shift may

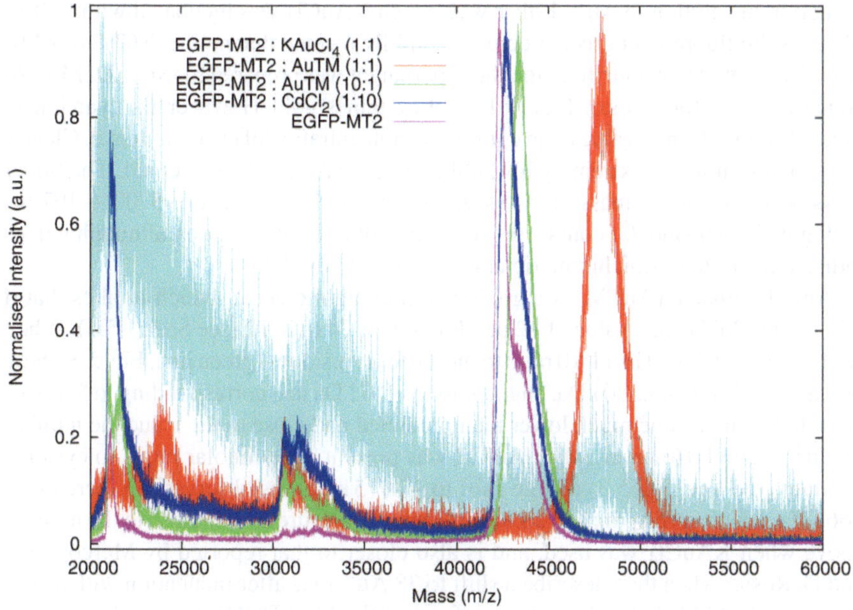

Fig. 5.3 MALDI MS of EGFP-MT2 after incubation with heavy-metal salts. Numbers in parentheses are molar ratios of cysteine present within MT2 to metal compound. The concentration of EGFP-MT2 is always 50 μM, which equates to 2 mM cysteine since there are 40 cysteine residues per protein molecule. The maxima of the peaks for each compound were set to unity

represent ($297 \div 58.69 =$) 5.1 bound Ni^{2+} ions. There is also a shoulder present at approximately 43.5 kDa, which may indicate 23 Zn^{2+} or 25.6 Ni^{2+} bound. These conclusions are questionable since MT binds only 7 divalent ions (see Sect. 1.5.4), so MT2 should bind to 14 divalent ions. It may be that the MT2 was partially oxidised, reducing the capacity of the domain to bind metal. This was previously shown to reduce platinum binding by ~50 % [6]. None of the buffers used were degassed, since the binding of metals *in vivo* was unknown and the reducing environment of the cell may prevent disulphide bridges forming. Nonetheless, the value of 42.309 kDa is a useful baseline to calculate the numbers of Cd^{2+} or Au^{+} bound.

To bind metal ions, EGFP-MT2 was incubated with various concentrations of heavy metal compounds at 37 °C for 3 h. Samples incubated with a 10:1 excess of $CdCl_2$ to cysteine residues gave a clear peak at 42.819 kDa, an increase of 510 Da (blue line, Fig. 5.3). This corresponded to 4.5 Cd^{2+} ions bound. This is much smaller than the 14 ions expected if Cd^{2+} bound with a ratio of 7 per MT domain, although it is curiously close to the number of Zn^{2+} ions bound if the above conclusion is correct. This diminished binding could also be explained by the acidifying effect of the MALDI matrix; acidification of MT is used to release Zn^{2+} during purification procedures [14] and may have a similar effect on cadmium binding. The small peak shift could also be explained by MT2 misfolding or forming disulphide bridges, as described above.

$KAuCl_4$ may donate Au^{3+} ions to MT, or alternatively may undergo spontaneous reduction in solution to Au^{+}. Either way, when $KAuCl_4$ was incubated with EGFP-MT2 a solid fluorescent green precipitate rapidly formed that neither TCEP or βME would disrupt. After centrifugation the supernatant was prepared for MALDI MS. No signal was seen for ratios of 10:1 or 1:10 of cysteines:gold. However, the cyan line in Fig. 5.3 shows the m/z value of an equimolar concentration of cysteine:metal. Clearly the signal to noise is extremely low, although 3 peaks can be discerned. The upper mass peak indicates a mass of ~48 kDa, which would represent ($\sim 5700 \div 197 =$) ~29 gold ions bound. Of course, the resolution of this value is poor, although it does indicate better than stoichiometric amounts of gold binding.

Aurothiomalate (AuTM) is a gold(I) containing molecule, which implies that it will bind to MT with a value of 12 per domain, or 24 to MT2 (see Sect. 1.5.4). When AuTM was incubated in a 1:10 ratio to the cysteine residues (green line, Fig. 5.3) there was a peak shift to 43.380 kDa, an increase of 1071 Da and corresponding to 5.4 Au^{+} bound. Again, a value much lower than predicted and close to the value the number of Cd^{2+} bound. However, when AuTM was present at equimolar ratios to cysteine residues (red line, Fig. 5.3) a large shift to 47.975 kDa was observed, an increase of 5666 Da, which corresponds to 28.8 Au^{+} ions. This figure is very close to the number above when $KAuCl_4$ was used, and is also closer to that reported by Mercogliano and DeRosier when they describe a shift to 38 Au^{+} ions after incubation with a 10:1 excess of AuTM [1]. The Fourier transform of the MALDI MS trace did not reveal a periodicity around 197 Da (data not shown), which would be expected from single Au^{+} ions binding to EGFP-MT2. Nonetheless, it is likely that this number does reflect single Au^{+} ions binding to EGFP-MT2, with no thiomalate moiety bound, as previously reported [15].

Electrospray ionisation mass spectrometry (ESI MS) was performed on samples incubated with an equimolar ratio of AuTM. However, the resulting spectrum had a low signal-to-noise ratio and a very broad peak. This was probably caused by the wide distribution of Au^+ ions bound combined with the multiple charges present on samples subjected to ESI MS.

Mass spectrometry demonstrated the metal binding capabilities of EGFP-MT2. Cadmium showed poor binding in all the buffers tested. Indeed, spontaneous reduction to cadmium colloids occurred in buffers containing morpholino rings (e.g., MOPS) and $CdCl_2$ showed a propensity for precipitating some of the protein out of solution. The purification tag, His_6, was not cleaved off for these experiments and may either bind to incubated metal or have leached some from the nickel column. However, this is not a major concern since His_6 only binds to a maximum of 2 nickel ions so, in the case of the gold incubation, would only affect the result by 2 %.

Despite $KAuCl_4$ showing very poor signal-to-noise, it also showed a clear peak shift at relatively low (2 mM) concentration. AuTM also showed a clear increase in mass, with both gold compounds binding to ~29 gold ions. For this reason, imaging focussed mainly on these two compounds.

5.1.4 Dynamic Light Scattering

Dynamic light scattering (DLS) can be used to determine the hydrodynamic diameter of small particles in solution. By measuring how the back-scattered light from a monochromatic laser varies with time, the diffusion coefficient of the particles can be measured, from which the hydrodynamic diameter of the protein can be calculated.

The dimensions of GFP are approximately 3 × 5 nm, whilst DLS of His_6-EGFP gave a hydrodynamic diameter of 6 nm (not shown). This slight discrepancy may be caused by the shell of water molecules surrounding the protein as well as the additional His_6 tag, and confirms the validity of this technique to measure small proteins in solution.

DLS of His_6-EGFP-MT2 gave a hydrodynamic diameter of 13.5 nm (Fig. 5.4, green line). Each individual MT domain is known to be approximately 2 × 4 nm [16]. When concatenated, however, the folding of the domain, and therefore its size, was unknown. This diameter may also have reflected some bound metal, either Zi^{2+} leached from the growth media or Ni^{2+} stripped from the purification column. Incubation with the metal-chelator EDTA increased the diameter, which varied between 13.5–16.6 nm (data not shown). This may represent MT reverting to a random coil, which may then have many conformations and most likely a more open structure than a tightly-folded MT domain. In agreement with this hypothesis, incubation of EGFP-MT2 with increasing concentrations of AuTM actually led to an overall decrease in hydrodynamic diameter (Fig. 5.4). In this case, 50 μM protein was incubated with 0.2, 2, or 20 mM AuTM, representing a 10:1, 1:1, or 1:10 ratio of cysteine:gold, respectively. These gave diameters of 8.7 nm for 0.2 mM AuTM, 11.7 nm for 2 mM AuTM, and 5.6 nm for 20 mM AuTM (Fig. 5.4, blue lines). The overall decrease

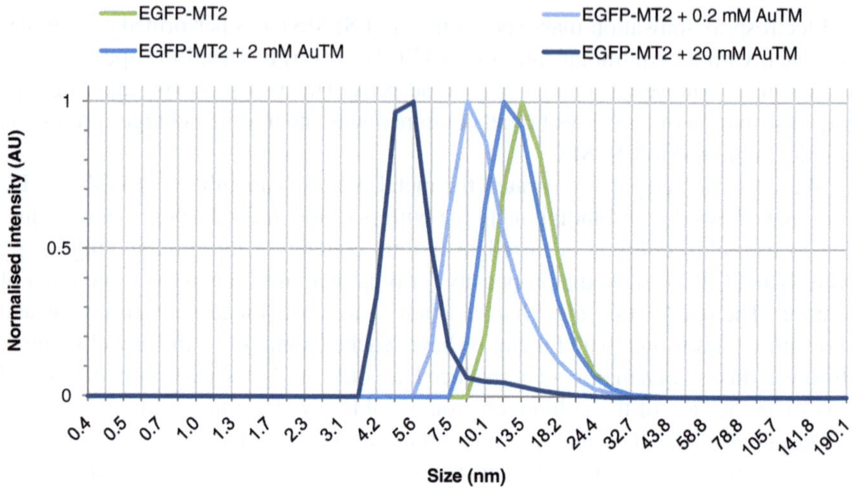

Fig. 5.4 DLS of EGFP-MT2 incubated with increasing concentrations of AuTM. Maxima were scaled to unity

in diameter may represent the MT2 forming a tighter coil around the gold cluster, resulting in a smaller diameter. If MT2 was clustering gold ions, the scattering properties of the complex may change to reflect this. Generally, inorganic particles backscatter more light than proteins. The DLS itself modulates the amount of light being backscattered with the use of an attenuator, which blocks light resulting in a fairly even amount of light reaching the detector. In this case, however, the attenuator stayed at the same value. This may mean that either nanoparticles are not forming, which is unlikely given the change in size of the protein, or the nanoparticles are too small to efficiently scatter light.

When incubated with 1 mM $CdCl_2$ there were two peaks with diameters of 12.1 and 74 nm (not shown). The peak at 12.1 nm overlaped closely with the native EGFP-MT2 whereas the peak at 74 nm may represent protein-mediated cadmium nanoparticles, although this was much larger than the 2 nm nanoparticles predicted. Similarly, when incubated with 10 mM $KAuCl_4$, the particle diameter increased to 251 nm. These were most likely to be precipitates of protein and metal, as described above.

5.2 Imaging EGFP-MT2 *In Vitro*

Having shown by MALDI MS that EGFP-MT2 binds to gold ions, the protein was targeted to specific areas for localisation in the TEM. EGFP-MT2 was encapsulated within lipid vesicles, or genetically fused to a protein that tubulates liposomes.

5.2.1 Localisation Within Liposomes

To localise EGFP-MT2 for visualisation by TEM, liposomes encapsulating EGFP or EGFP-MT2 were produced by sonication and extrusion before incubation with either $KAuCl_4$ or AuTM. Before metal incubation, liposomes were separated from non-encapsulated protein by ultracentrifugation and resuspension of the liposome pellet.

Liposomes prepared by air-drying were not stabilised by osmium or negatively stained with uranyl acetate. This was to increase the contrast from any gold particles formed. However, the liposomes exhibited extensive beam-damage in the electron microscope (Fig. 5.5). For this reason, liposomes were stabilised by methylcellulose with a small amount of uranyl acetate.

When treated with methylcellulose liposomes could be more easily located in the TEM, indicating decreased sublimation in the electron beam. However, the osmotic pressure was still too great and most liposomes were found to have ruptured. Liposomes encapsulating EGFP-MT2 that had not been incubated with metal compounds exhibited small particles (Fig. 5.6a, white arrowheads), however these were not uniform in size and much less electron-dense than those seen after incubation with $KAuCl_4$ (Fig. 5.6c, black arrowheads). They are likely to be lipid fragments that were not incorporated into liposomes, a conclusion reinforced by the non-specific distribution of the particles across the grid and exhibiting no localisation with liposomes.

Liposomes encapsulating EGFP-MT2 incubated with 0.1 mM $KAuCl_4$ were substantially more electron-dense than liposomes not exposed to gold (Fig. 5.6b), indicating that gold can passively cross lipid membranes and increase the electron density of encapsulating particles. In the vicinity of ruptured liposomes there were numerous electron-dense particles of approximately 2 nm in diameter (Fig. 5.6c, d, black arrowheads), the size expected from the MT2 domain [16]. This indicates that EGFP-MT2 free in solution can bind to $KAuCl_4$ and form electron dense particles clearly visible by TEM.

When incubated with AuTM, no nanoparticles were imaged around or inside the liposomes (not shown). This may be because the concentration of protein after encapsulation is difficult to ascertain, and as such the concentration of AuTM may not have been high enough.

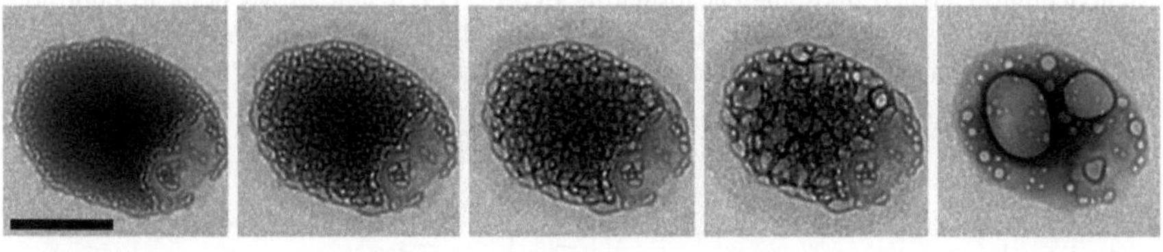

Fig. 5.5 A kymograph of electron-beam damage to vesicles encapsulating His_6-EGFP. Images were taken at 0, 15, 30, 60, and 900 s. Vesicles were negatively stained with uranyl acetate. Scale bar is 100 nm

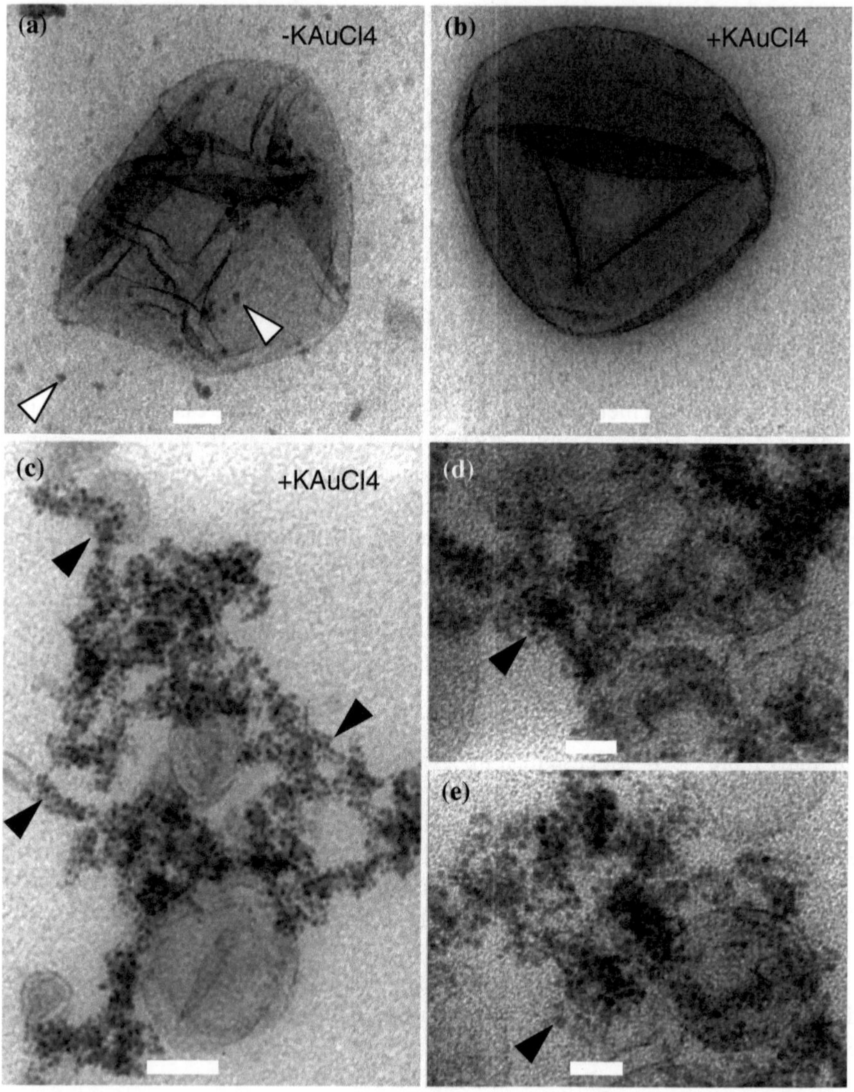

Fig. 5.6 Liposomes treated with methylcellulose incubated with or without $KAuCl_4$. **a** Methylcellulose-coated liposome encapsulating EGFP-MT2. *White* arrows point to faint non-specific particles. **b** and with 0.1 mM $KAuCl_4$. **c** Ruptured liposomes in the presence of 0.1 mM $KAuCl_4$. **d** and **e** Magnified images of nanoparticles. Scale bars are 50 nm for (**a**–**c**) and 20 nm for (**d** and **e**)

5.2.2 Membrane Tubules

Protein mediated membrane tubulation has been reported previously with various proteins, including amphiphysin and various sorting nexin (SNX) proteins [17]. *In vivo*, sorting nexin-1 (SNX1) mediates endosomal protein sorting through the retromer [18]. SNX proteins both sense and drive membrane curvature, producing tubular vesicles that are enriched with cargo protein [17]. However, there is little knowledge of how SNX proteins achieve membrane curvature and elongation. For this reason, SNX1 was chosen for tagging with EGFP-MT2, producing EGFP-MT2-SNX1, named EMS1 herein. This was purified by nickel-affinity chromatography as described above and revealed a band at the predicted weight of 102 kDa (Fig. 5.7).

Liposomes were formed from bovine brain lipid extracts supplemented with freshly protonated phosphatidylinositol 3-phosphate (PI(3)P) and extruded to 200 nm in diameter, which was confirmed by DLS. Purified EMS1 was incubated with AuTM (1:1 gold to cysteine residues) before being added to the liposomes and incubated at room temperature for 10 min. EMS1 incubated with $KAuCl_4$ formed solid fluorescent green precipitates that removed all tubulating propensity from the sample. The liposomes were stabilised with methylcellulose and either left unstained or negatively stained with uranyl acetate before viewing in the TEM (Fig. 5.8). EMS1-mediated tubules appeared as a "string of sausages", with constrictions spread evenly along the length of the tubule (Fig. 5.8c). Figure 5.8d shows a diffuse electron-dense particle or region between successive "sausages" that may be indicative of EMS1. Between each dark particle is a dark band, which may be EMS1 forming a helix around the tubule, as is the mechanism proposed [19, 20]. The banding pattern was consistent with all tubules imaged, although the particles were not (Fig. 5.8f). When no negative

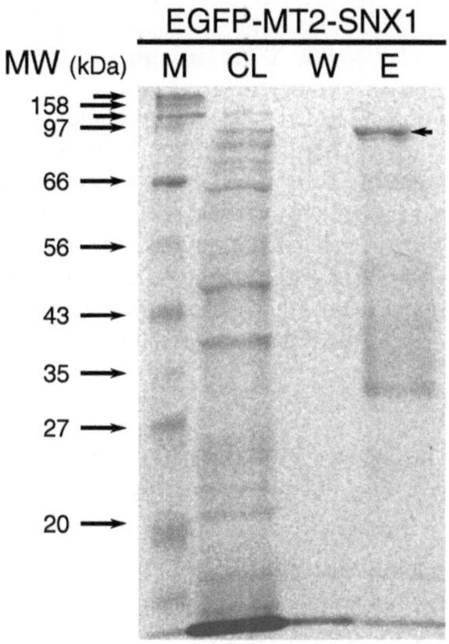

Fig. 5.7 Purification of EMS1. *Arrow* indicates purified EMS1 at 102.4 kDa. *M* Marker; *CL* Cleared lysate; *W* Wash; *E* Eluate

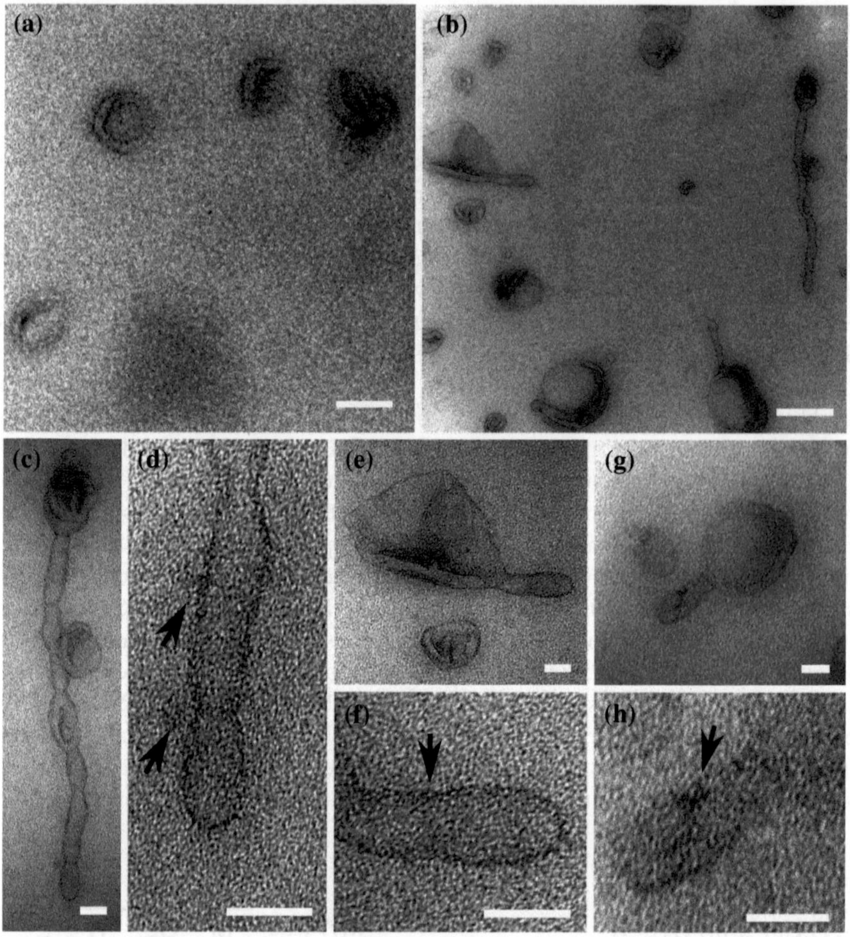

Fig. 5.8 EMS1 tubulates liposome membranes. **a** Liposomes without protein are monodisperse 200 nm spherical structures. **b** In the presence of EMS1 the membranes are distended to form tubules. **c** Tubules had the appearance of a "string of sausages", with wider portions spaced evenly along their length. **d** Magnifying (**c**) reveals diffuse dark particles at the concave edge of the thinner portions and a dark banding across the tubule, possibly indicating the presence of EMS1 (*arrows*). **e** and **f** The dark banding pattern was consistently visible (*arrow*), although the particles were not. **g** and **h** Imaging liposomes without negative stain revealed a helical pattern around the tubule (*arrow*). Scale bars for (**a** and **b**) are 200 nm, for (**c**–**h**) they are 50 nm

stain is used the dark banding pattern appears much clearer (Fig. 5.8 g, h), as would be expected when contrast is formed by Au^{+} alone and not UO_2^{2+}.

Previously reported mechanisms have indicated that the SNX tubule is densely packed around the liposome core [19]. This hypothesis, however, leaves little room for transmembrane proteins that are known to bind cargo [21]. If these images faithfully represent the SNX1 helix it appears that the tubulating coat is much less densely packed than previously thought. CryoTEM of liposomes has been performed in an attempt to distinguish between the protein coat and lipid bilayer. Unfortunately, the efficiency of tubulation by EMS1 was not high enough to reliably image the protein-mediated tubules. Tubulation is highly dependent on lipid composition; indeed, several batches of bovine brain lipid from the same supplier did not yield any tubules at all, and lipids that did were rapidly oxidised despite storage under nitrogen. It may be that blending lipids in exact, and known, compositions would allow the efficiency of tubulation to be high enough to faithfully image tubules by cryoTEM.

5.3 Imaging EGFP-MT2 *In Vivo*

Concatenated MT was required to increase clustering of metal ions to form electron-dense nanoparticles. However a tandem gene fusion containing two identical sequences is undesirable due to potential homologous recombination and secondary structure elements interfering with PCR. For this reason, a non-identical MT2 domain was designed using the program Gene Designer. Computationally, the protein sequence used for *in vitro* studies was back-translated to DNA with the following restrictions; all repeat sequences longer than 10 bp were removed, as well as all of the RE sites also found in the vector multiple cloning site (MCS). RE sites that would be used for subcloning were added and modified to allow easy removal, addition of extra proteins, or further concatenations of MT. Codon optimisation was performed for human cell expression [22] before the insert was synthesised and subcloned. This yielded a vector that was compatible with in-frame insertions before, after, and in between EGFP-MT2.

5.3.1 CLEM of Intracellular Fusion Proteins

Imaging EGFP-MT2 in both the light and electron microscope was central to this project and a critical attribute for a probe designed for CLEM. EGFP-MT2 could be imaged in the light microscope with ease, since it contains brightly-fluorescent EGFP. However, in order to trace back the same structure imaged in the LM it was crucial to localise the probe to a structure easily identifiable in the TEM. For this reason, EGFP-MT2 was subcloned next to Sec23A, SNX1, α-Tubulin, the Transferrin receptor (TfnR), and Tetherin (Fig. 5.9); proteins which localise to specific regions of the cell that are identifiable in the TEM. Sec23A marks regions of ER-exit sites and COPII-coated vesicles [23], SNX1 forms tubules along recycling endosomes [21], α-tubulin is a component of the microtubule cytoskeleton [24], the transferrin receptor cycles between the cell surface and early endosome [25], and Tetherin cycles between the trans-Golgi network and plasma membrane [26].

Cells were incubated with either $KAuCl_4$ or AuTM, not K_2PtCl_4 or $CdCl_2$. This was because K_2PtCl_4 showed no binding by MALDI-MS and was found to

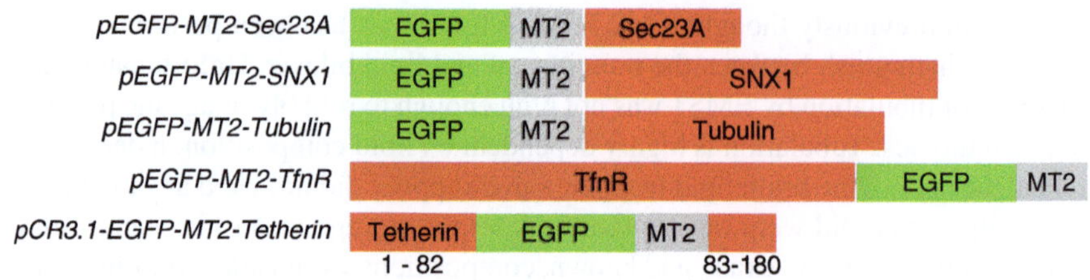

Fig. 5.9 A schematic showing the protein constructs used in this section. The length of each domain corresponds to its molecular weight. Residue numbers are shown for the Tetherin construct, where EGFP-MT2 is inserted within the Tetherin protein

spontaneously reduce in the buffers used for cell culture (data not shown). $CdCl_2$ was not tried because it also showed minimal binding in MALDI-MS and proved to be highly toxic to cells. Despite showing a high increase in apoptosis above 30 μM using cell apoptosis assays, a dose of 10 μM was enough to completely alter the morphology of the cells, as judged by LM. Since this is designed to be a general probe for CLEM, cadmium was not used.

5.3.1.1 Tubulin

α-tubulin forms a dimer with β-tubulin that assemble to form microtubules. EGFP-tubulin is routinely used to study microtubule dynamics. However, unlike EGFP-tagged tubulin, addition of the EGFP-MT2 domain caused significant disruption to the structure to render the tubulin fragment non-functional; no fluorescent microtubules were imaged. Even when cells expressing EGFP-MT2-tubulin were treated with Taxol, a microtubule-stabilising drug [27], only cytoplasmic fluorescence was seen. For this reason, future work on EGFP-MT2-tubulin was abandoned and other intracellular proteins were tagged.

5.3.1.2 Sec23A

To elucidate whether the MT2 domain disrupts cellular function, HeLa cells transfected with pEGFP-MT2-Sec23A were immunolabelled with antibodies to various cellular proteins (Fig. 5.10). Sec23A is a component of the COPII complex, which mediates anterograde transport from ER exit sites (ERES) to the Golgi apparatus [23]. Sec23A acts as a Sar1 GAP, a function that is not perturbed by fusion to EGFP [28, 29]. It was important, however, to determine whether the addition of the cysteine-rich MT2 domain adversely affected the structure or function of Sec23A.

Sec23A forms a heterodimer with Sec24C, which is involved in cargo-binding and forms the prebudding complex [23]. Fusion to EGFP-MT2 did not disrupt formation of the prebudding complex, as EGFP-MT2-Sec23A and Sec24 colocalised

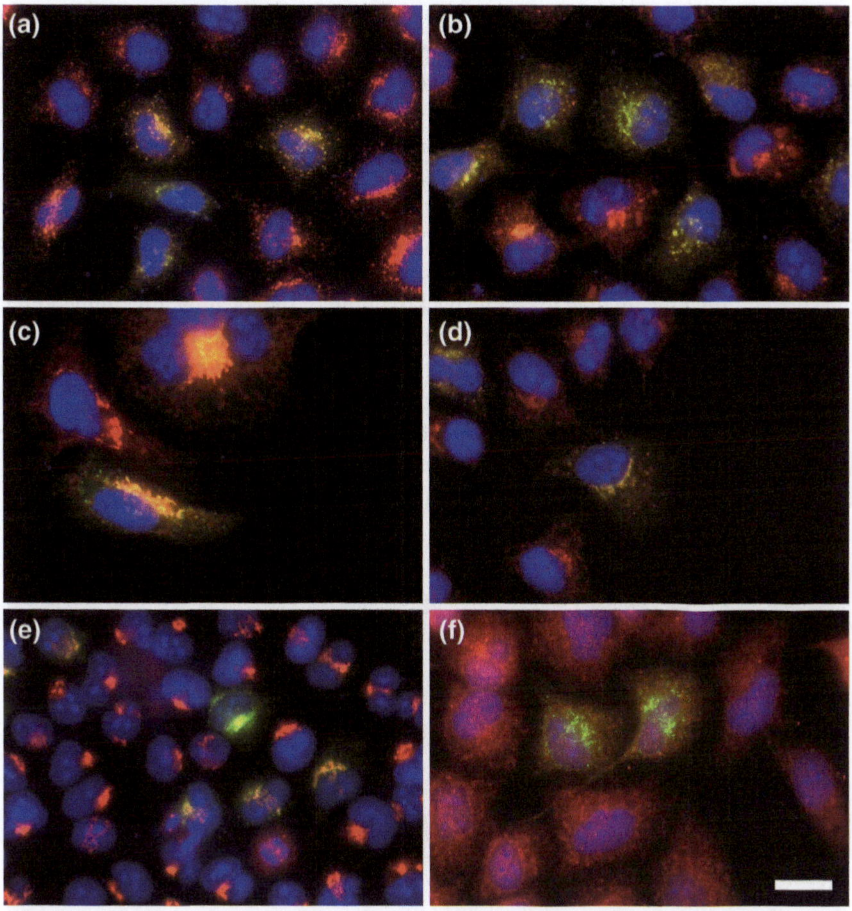

Fig. 5.10 Immunofluorescence of HeLa cells expressing EGFP-MT2-Sec23A. HeLa cells expressing EGFP-MT2-Sec23A (*green*) were immunolabelled (*red*) against Sec24C (**a**), Sec31A (**b**), β'-COP (**c**), GM130 (**d**), Giantin **e**, and Calnexin (**f**). The nucleus was counterstained with DAPI (*blue*). Scale bar is 20 μm for all images

(Fig. 5.10a). Two each of Sec31A and Sec13 form a tetramer that binds to the outside of the Sec23/24 dimer [23]. Sec13/31A may serve as an analogous function to clathrin in generation and maintenance of membrane curvature [30]. Sec31A was shown to colocalise with Sec23A (Fig. 5.10b), illustrating that the addition of the MT2 domain did not affect coat formation. β'-COP is a component of the coatomer complex of the COPI coat [31]. COPI functions in the opposite direction to COPII, mediating retrograde transport from the Golgi apparatus to ER, which is dependent on anterograde transport by COPII. Discrete vesicles were distinguished in Fig. 5.10c, implying functional retrograde transport. GM130 is localised to the cis-Golgi [32] where it mediates cisternal stacking [33] and possibly also vesicle fusion [34]. Giantin has a structural function, mediating Golgi stacking and vesicle fusion [35]. Any disruption to cargo binding or transport would disrupt the morphology of the Golgi stacks. As such, Fig. 5.10d, e showed normal cellular localisation and structure of the Golgi. Calnexin is a chaperone protein retained in the endoplasmic reticulum (ER) [36]. ERES can be distinguished as discrete loci on the ER network in Fig. 5.10f, again demonstrating unperturbed function of Sec23A.

Since the addition of EGFP-MT2 did not perturb the structure or function of Sec23A, samples were prepared for CLEM. The procedure followed the scheme outlined in Sect. 3.2. Firstly, a cell expressing EGFP-MT2-Sec23A was located and the surrounding area imaged to enable backtracking of the location after embedding. The cell was then imaged as a timecourse to ensure protein trafficking was not perturbed. Finally, prior to chemical fixation, a z-stack of the cell was acquired to allow determination of the height of the serial section imaged in the TEM.

The cells shown in Fig. 5.11 were incubated with 0.1 mM $KAuCl_4$ for 3 h before fixation in 2.5 % glutaraldehyde and embedding in Epon. The cell was traced back using the finder grid and sectioned for TEM. Overlaying the fluorescence image

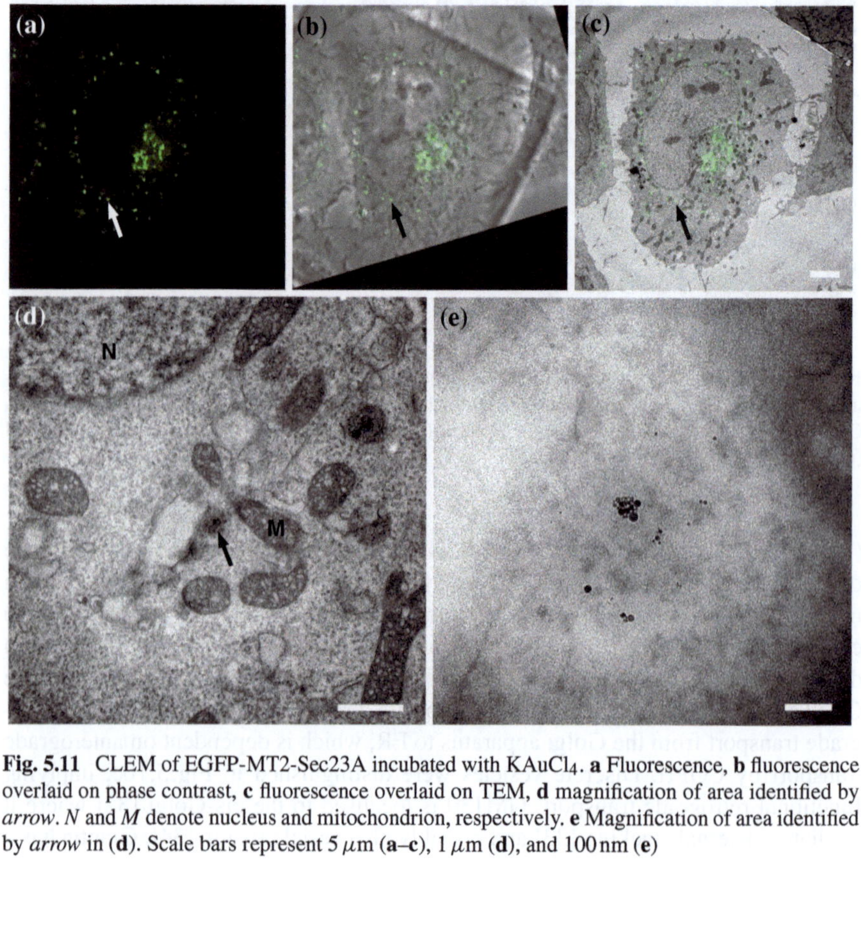

Fig. 5.11 CLEM of EGFP-MT2-Sec23A incubated with $KAuCl_4$. **a** Fluorescence, **b** fluorescence overlaid on phase contrast, **c** fluorescence overlaid on TEM, **d** magnification of area identified by *arrow*. *N* and *M* denote nucleus and mitochondrion, respectively. **e** Magnification of area identified by *arrow* in (**d**). Scale bars represent 5 μm (**a**–**c**), 1 μm (**d**), and 100 nm (**e**)

allowed the same structure/organelle to be found in the TEM (arrow in Fig. 5.11). In this case the fluorescence was localised to a membrane enclosed sac, possibly part of the Golgi or a COPII-coated vesicle but impossible to tell without imaging serial slices or TEM tomography. Magnifying the area yielded electron-dense nanoparticles close to the membrane (Fig. 5.11e).

As shown in Fig. 5.11, nanoparticles were visible in the same location as fluorescence (within the resolution of the LM). However, these particles were rare and, although they were probably MT2-mediated nanoparticles, immunolabelling is essential to prove this. Unfortunately, it was also difficult to identify ERES and COPII-coated vesicles in both transfected and control cells, so instead a more recognisable intracellular domain was used; the early endosome.

5.3.1.3 SNX1

Sorting nexin-1 (SNX1), described above (Sect. 5.2.2), mediates vesicular transport by cargo sorting at the endosome and tubule budding from the retromer complex [18]. SNX1 tagged to EGFP-MT2 displayed dynamic tubular structures *in vivo* identical to EGFP-tagged SNX1 [17], although fluorescent aggregates could be seen in cells overexpressing the construct. For this reason, only cells that displayed clear tubular vesicles and functioning transport were considered for processing for TEM.

After incubation with $KAuCl_4$ or AuTM, cells expressing EGFP-MT2-SNX1 displayed clear endosomes. Processing for CLEM allowed tubules emanating from endosome to be clearly visualised (Fig. 5.12). Unfortunately, the endosomes did not show clear labelling of any helical array. Although there is a faint electron dense "smear" along the right of the tubule in Fig. 5.12c this was not clear or reliably reproducible.

It may be that the nanoparticles formed were too small to be visualised clearly with standard TEM techniques. This was why proteins were chosen that polymerised;

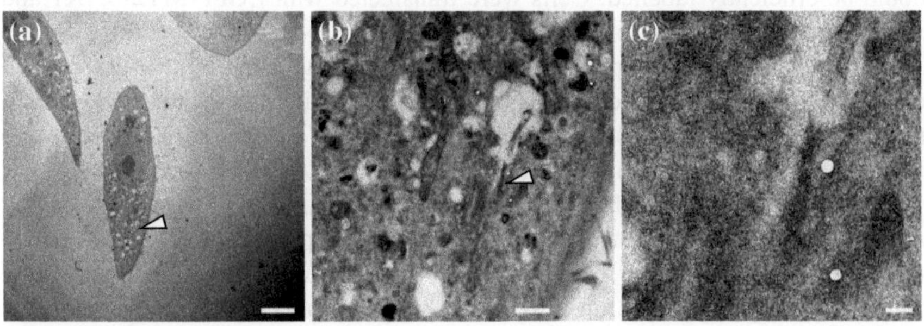

Fig. 5.12 CLEM of EGFP-MT2-SNX1 incubated with $KAuCl_4$. **a** Fluorescence overlaid on TEM, with a fluorescently-labelled endosome (*arrow*). **b** Magnification of area identified with tubule highlighted (*arrow*). **c** Tubule emanating from endosome. Scale bars are 10 μm, 1 μm and 200 nm from *left to right*

Sec23A around ERES and SNX1 around tubules. However, since this signal was still ambiguous it was decided to increase the likelihood of seeing MT2-mediated nanoparticles by enhancing the signal with autometallography, specifically by gold deposition.

Gold deposition onto gold nanoparticles is already used to enhance signals for LM and protein gels, and has recently been applied to TEM studies. Commercial gold enhancement kits are available but in this case gave very high background. For this reason a modified protocol based on that developed by Morphew et al. was used [37], which slowed down development time and decreased background labelling.

Figure 5.13 clearly shows a non-transfected cell to the left of a transfected cell. Although in the transfected cell dark particles could be clearly seen surrounding a tubule emanating from an endosome (Fig. 5.13c), in the non-transfected cell there were also dark particles (Fig. 5.13b). The particles at the left of (b) appear to be within a late endosome or lysosome, perhaps indicating that $KAuCl_4$ spontaneously reduces in DMEM, and it is these particles which are endocytosed by both transfected and non-transfected cells alike. An added wash step after the 3 h incubation with heavy metal compounds may allow the cell to exocytose any engorged gold particles.

Because it appeared that $KAuCl_4$ was reduced by the buffers or within the cells, research focussed on AuTM. Since AuTM treated cells did not show nanoparticles either before or after gold enhancement, in either transfected or non-transfected cells (not shown) it was decided to localise gold-labelled antibodies to EGFP in AuTM treated cells to localise tagged proteins at high resolution. The Tokuyasu method was developed to section cells without plastic embedding [38]. This also preserves the antigenicity of the sample, enabling immunolabelling of cells for TEM with a much higher efficiency than plastic embedded cells [39]. In contrast to plastic embedded cells, where labelling occurs only at the surface of the section, Tokuyasu immunolabelling penetrates deeper into the section. Although this technique can be used to perform CLEM [39], in this case only ultrastructural localisation was desired so no LM was performed. Since the antibody was to GFP, non-transfected cells would show only minimal background binding (preferably none), whereas transfected cells should be efficiently labelled. Cells were transfected with EGFP-MT2-SNX1 and imaged to ensure transfection efficiency (in this case, $\sim$50 %) before being lightly fixed. The cells were then pelleted by embedding in gelatin, gently scraped and centrifuged. The resulting pellet was hardened by cooling the gelatin and cut into small cubes ($\sim$1 mm^3). These were infused with a high concentration of sucrose to prevent ice crystal formation before being frozen in $N_2(l)$ and cryosectioned.

Immunolabelling was performed with anti-GFP antibodies that were then labelled with protein-A (PA) conjugated to 10 nm colloidal gold particles. PA binds to the Fc domain of the antibodies of many species in a 1:1 ratio. This is in contrast to IgG immunolabelling where IgG binds to many epitopes along each antibody and therefore results in greater than 1:1 binding, a feature that can be useful for low numbers of antigenic sites as it effectively amplifies the signal in TEM. As shown in Fig. 5.14, labelling was efficient and clearly shows membrane tubules and endosomes. Membranes appear different after preparation for the Tokuyasu method due to negative staining: For plastic embedding, OsO_4 is added to stabilise lipid membranes and

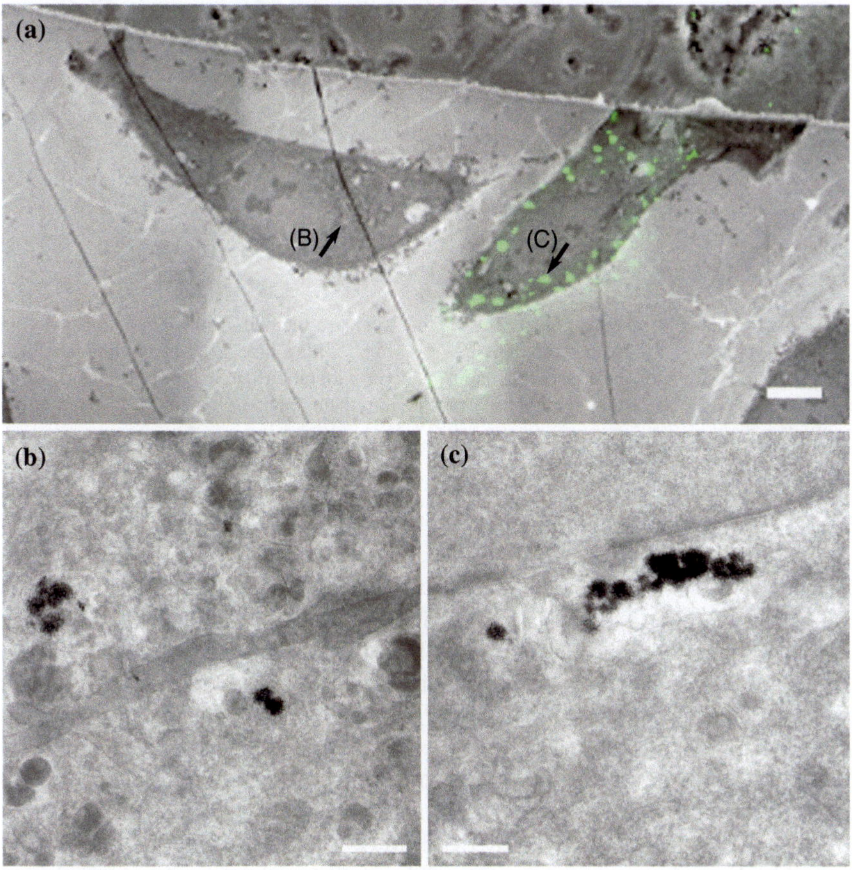

Fig. 5.13 Gold enhancement of EGFP-MT2-SNX1 incubated with $KAuCl_4$. **a** Fluorescence overlaid on TEM, showing both a transfected cell with a fluorescently-labelled endosome (*arrow*) and a non-transfected cell. Phase contrast is also visible (*upper right*), which was used to aid alignment of the TEM and FLM images. **b** Magnification of non-transfected cell with the area identified with the *arrow* in (**a**), clearly showing gold particles within and around membrane sacs. **c** Transfected cell area identified with the arrow in (**a**) magnified, showing gold particles surrounding a membrane tubule. Scale bars are 10 μm for (**a**) and 500 nm for (**b**) and (**c**)

therefore acts as a positive stain, making lipids appear dark. For Tokuyasu, however, uranyl acetate acts as a negative stain for membranes and they therefore appear as white belts surrounded by stain.

In AuTM treated cells there did not appear to be any nanoparticles co-localised with the 10 nm gold. This may be because the preparation procedure actually washes out any protein-bound particles. Due to the difficulty in localising MT2 intracellularly it was decided to move to an extracellular protein marker: Specifically, the receptor for transferrin.

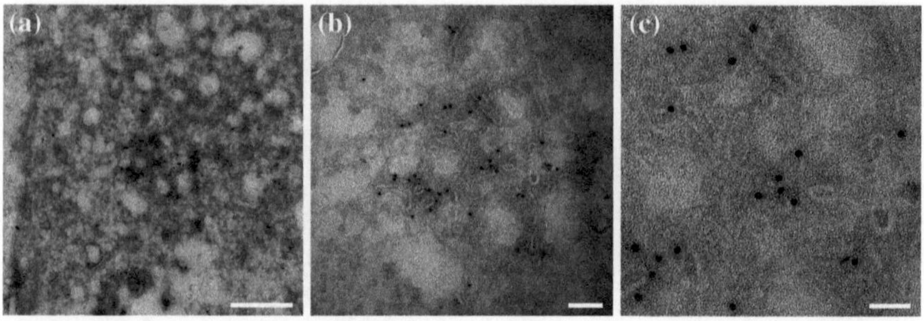

Fig. 5.14 ImmunoEM on cells prepared using the Tokuyasu method. Cells were incubated with 0.1 mM AuTM for 3 h prior to fixation. **a** Region of cell with a high level of immunolabelling. **b** Magnification of region in (**a**) showing endosome sacs with tubules. **c** Magnified image showing colloidal gold particles, but no extra density proximal to them. Scale bars are 500, 100 and 50 nm from *left to right*

5.3.1.4 Transferrin Receptor

Transferrin receptor (TfR) is a homodimer that cycles between the plasma membrane and early endosomes, where its transferrin cargo releases iron in a pH dependent manner [25]. Intracellular proteins were tried first because: (1) methods exist for labelling proteins that appear extracellularly, and (2) the extracellular environment is generally oxidising, as opposed to the reducing environment within the cell. It was not known how well the 40 cysteine residues would resist disulphide bridge formation. However, disulphide bridge formation is resisted when metal ions are bound to MT [40], thus it may be that the added gold prevents oxidation.

When cells transfected with TfR-EGFP-MT2 were treated with 0.1 mM AuTM and gold enhanced, nanoparticles were visible in both transfected and non-transfected cells (Fig. 5.15). They were also present in late endosomes/lysosomes, which may indicate mis-trafficking of TfR, possibly due to oxidation of the MT2 domain causing disulphide bridge formation. Mis-trafficking of TfR is also a problem when transferrin itself is conjugated to QDs [41], indicating a highly controlled pathway.

Diestra et al. had imaged MT3 probes *in vivo* when using AuCl [10, 11], a compound with very low solubility in aqueous solutions, it is likely that they were in fact imaging particles formed following disproportionation. These could be formed from either $AuCl_4^-$ or even gold colloids formed from Au^0. To test this hypothesis, gold nanoparticles (AuNPs) were formed from chloroauric acid ($HAuCl_4$) reduced with sodium borohydride ($NaBH_4$) and trisodium citrate. Characterisation by UV-vis spectrometry and DLS gave an average diameter of ~4 nm, which was confirmed by TEM.

Incubation of AuNPs with TfR-EGFP-MT2 transfected cells was followed with imaging using scanning TEM (STEM). STEM generates contrast based on the atomic number (Z) of the sample, which the beam rasters across with a resolution of ~2 nm. STEM has shown great contrast improvements when imaging AuNPs over conventional TEM (see Ref. [42], Fig. 5.8). However, with TfR no AuNPs were visible at

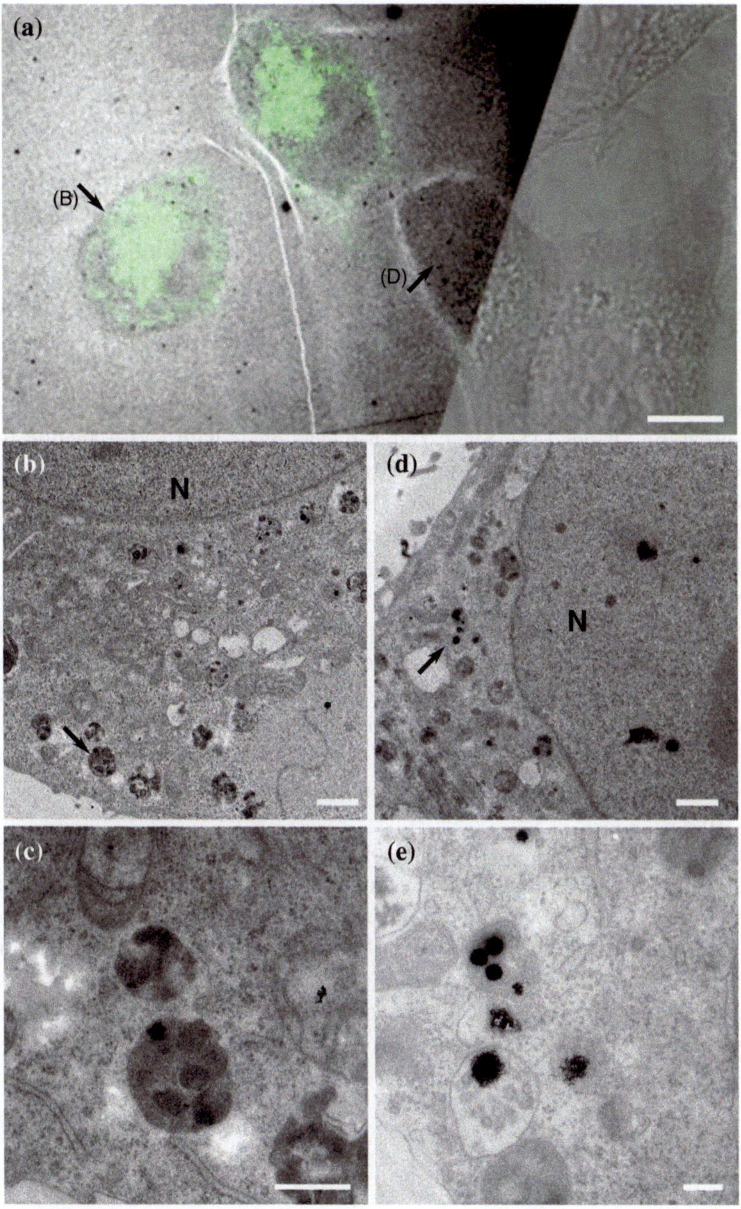

Fig. 5.15 CLEM of EGFP-MT2-TfR incubated with AuTM after gold enhancement. **a** Fluorescence overlaid on TEM, with phase contrast visible at the *right*. Note two transfected cells adjacent to one non-transfected cell. **b** Magnification of area of transfected cell identified by arrow. *N* denotes nucleus. **c** Magnification of area identified by *arrow* in (b), showing particles in a late endosome or lysosome. **d** Magnification of area of non-transfected cell identified by *arrow* in **a**. *N* denotes nucleus. **e** Magnification of area identified by *arrow* in **d**, showing particles in a late endosome or lysosome. Scale bars represent 10 μm (**a**), 1 μm (**b**), 500 nm (**c**), 1 μm (**d**), 200 nm (**e**)

any of the locations derived from FLM. This may be because MT2 will not bind to AuNPs this large, or perhaps oxidation reduced the binding capacity of MT2.

5.3.1.5 Tetherin

Tetherin cycles between the trans-Golgi network and plasma membrane [26]. Tetherin is a transmembrane protein that forms homodimers *via* formation of a parallel dimeric coiled coil and terminates at the C-terminus with a GPI anchor [43]. Tetherin has recently become an object of intense scrutiny based on its suspected role in inhibiting HIV budding and release, and therefore may have a pathological role in AIDS resistance [44]. Indeed, cells expressing large amount of Tetherin form strings of virions that are unable to bud completely, indicating that Tetherin forms a direct link to escaping virions [45]. Since Tetherin terminates with a GPI anchor, EGFP-MT2 was inserted into a known loop region of the protein between residues 82 and 83, just before the coiled coil region [43]. This internal fusion was well tolerated, and tagged Tetherin was trafficked to the plasma membrane as normal.

Transfected cells were incubated with $KAuCl_4$, AuTM or no gold and fixed with paraformaldehyde. This was followed by a gold enhancement step before immunolabelling for Tetherin. Cells with no gold incubation displayed no background nucleation of gold particles, indicating the modified enhancement reaction used to reduce background staining proceeded with high fidelity. Immunolabelling showed membrane-proximal colloidal gold particles as expected. However, cells that had been incubated with $KAuCl_4$ showed a similar result to the previous systems, with non-specific deposition of gold on both transfected and non-transfected cells (Fig. 5.16).

Incubation of cells with AuTM gave some more promising results (Fig. 5.17). In contrast to the previous samples, large aggregates of gold were not observed. Rather, nanoparticles were found clustered around membrane sacs in the cytoplasm and at the cell surface (Fig. 5.17b, c). Although there was no colocalisation with colloidal gold, there was a population of nanoparticles proximal to an immunolabelled site (Fig. 5.17c). Unfortunately, with this image there were no non-transfected cells to treat as negative controls.

5.4 Summary of Results

Characterisation of EGFP-MT2 by MALDI MS indicates that the MT2 domain is able to bind to ~29 Au^+ ions *in vitro*. This is considerably less than the 55–70 atoms found in 1.4 nm nanogold [15], although this could be visible if clustered around a substrate. For this reason, SNX1 was chosen as an *in vitro* localisation tag. EGFP-MT2-SNX1 (EMS1) was able to form membrane tubules from liposomes formed from bovine brain extract. It also showed signs of electron dense material bound along the length of the tubule, and imaging without stain revealed a dark

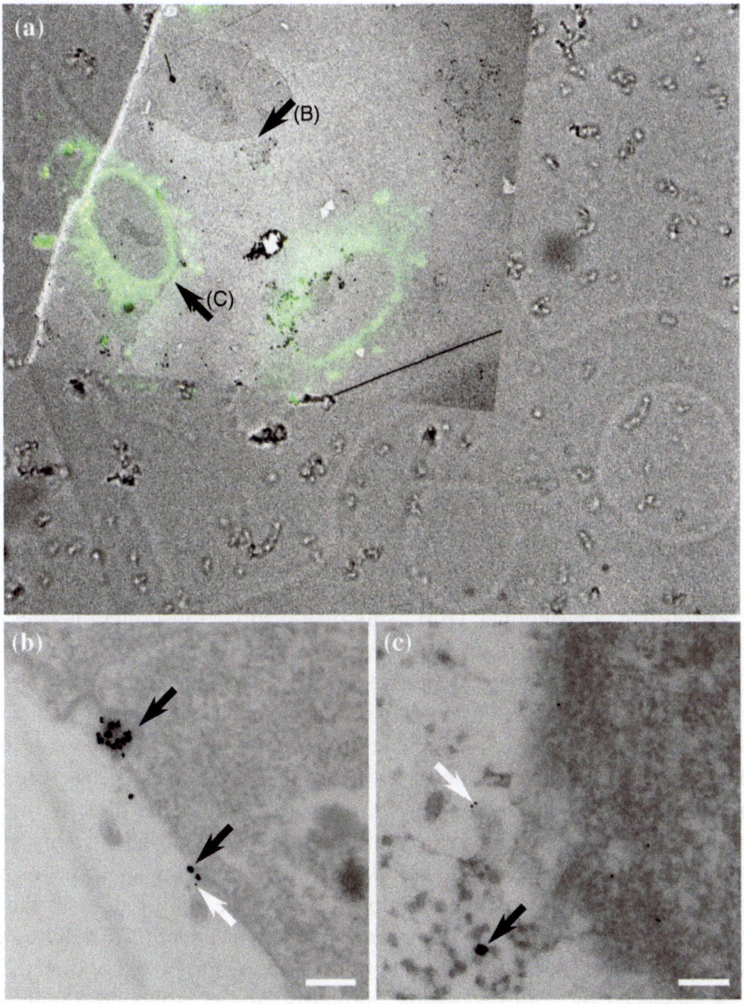

Fig. 5.16 CLEM of Tetherin-EGFP-MT2 incubated with KAuCl4 followed by gold enhancement. **a** Fluorescence overlaid on TEM and phase contrast image. Note two transfected cells adjacent to one non-transfected cell, as well as the number 8 of the CelLocate grid used to trace back cells for trimming. **b** Magnification of area of non-transfected cell identified by *arrow*. *White arrows* indicate gold colloids conjugated to antibodies, whereas *black arrows* indicate gold enhanced products (based on the size and shape of gold particles). **c** Magnification of area of transfected cell identified by *arrow*. *White arrows* indicate gold colloids conjugated to antibodies, whereas *black arrows* indicate gold enhanced products. Scale bars are both 200 nm

stripe across the tubule. This may be formed from gold particles conferring contrast to the sample. However, the tubulation assay was unreliable and time consuming. Optimisation of the assay should allow tubulated liposomes to be frozen and imaged by cryoTEM, which would allow any contrast generated by MT2-bound gold to show

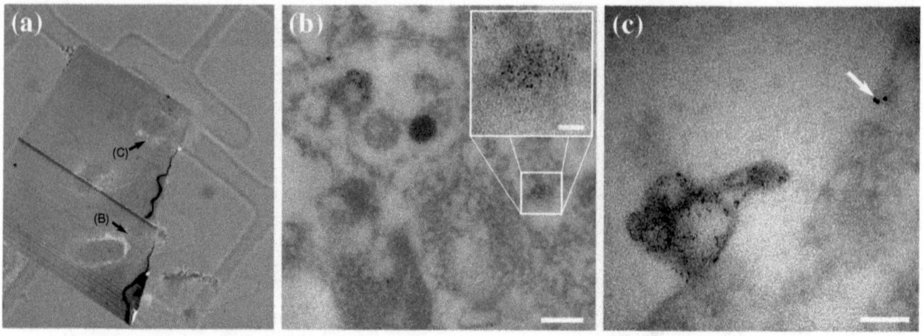

Fig. 5.17 CLEM of Tetherin-EGFP-MT2 incubated with AuTM after gold enhancement. **a** Fluorescence overlaid on TEM and phase contrast image. All the cells are transfected in this image, so there can be no internal control. The letter P is just visible behind the TEM overlay. **b** Magnification of area of transfected cell as identified by *arrow* in (**a**). Large clusters of gold-enhanced products can be seen intracellularly within membrane-enclosed sacs (inset). **c** Magnification as identified by *arrow* in (**a**). Three gold colloids conjugated to antibodies can be seen in the *upper right corner*, proximal to gold enhanced region surrounding a membrane. Scale bars are 100 nm for (**b**) and (**c**) and 20 nm for insert in (**b**)

more clearly than conventional TEM. This may also lead to protein helices being imaged and processed for helical reconstruction, allowing the SNX1 polymer to be imaged directly and the mode of action deduced.

For imaging EGFP-MT2 *in vivo* many different conditions and processing methods were tried (Table 5.1). Simple incubation of cells with gold compounds did not reliably show any clear nanoparticles, although there was one success (Fig. 5.11). Autometallography of gold nanoparticles that were not visible directly did yield many particles, although they were seemingly randomly distributed. This implied that, certainly for $KAuCl_4$, the gold compound was undergoing spontaneous reduction to solid gold, which was either endocytosed by the cell or stuck to the outer leaflet of the plasma membrane. A common theme was the inability to deliver enough gold to the MT2 domain for it to be able to bind enough to be either directly visualised or enhanced by gold deposition. Similarly AuTM, although a much more stable compound than $KAuCl_4$, was never visible directly. After gold enhancement of extracellular tagged proteins there were some notable results. Most obviously, EGFP-MT2-tagged Tetherin appeared to have discrete nanoparticles proximal to antibody-bound gold colloids. Unfortunately, this experiment had no internal control as previous experiments did (i.e., non-transfected cells incubated with AuTM were not visible in the TEM).

It appears that the MT-derived tags are unsuitable as general purpose protein labels for CLEM. *In vitro*, MT may increase the contrast of particles [3], even those expressed intracellularly and purified with metals attached. However, *in vivo* results have been few and far between. Studies using prokaryotes, which can withstand much higher concentrations of heavy metals, have shown some success [10, 11], but those performed in eukaryotic systems have shown both low resolution staining and extreme morphological changes due to heavy metal cytotoxicity [4]. The fact that

Table 5.1 Summary of results from *in vivo* CLEM

Protein tagged	Method	Condition	Nanoparticles?
Sec23A	Chemical fixation	$KAuCl_4$	✓
	Chemical fixation	AuTM	×
	Gold enhancement	AuTM	×
SNX1	Chemical fixation	$KAuCl_4$	✓
	Chemical fixation	AuTM	×
	Gold enhancement	$KAuCl_4$	✓
	Gold enhancement	AuTM	×
	Tokuyasu	AuTM	×
TfR	Gold enhancement	AuTM	✓
	STEM	AuNPs	×
Tetherin	Gold enhancement	$KAuCl_4$	✓
	Gold enhancement	AuTM	✓

MT is endogenous to most organisms has also not been taken into consideration, and may lead to background staining if the conditions for nanoparticle formations can be optimised.

5.5 Targeted Intracellular Delivery of Nanogold using pH-Sensitive Polymersomes

Since the MT2 tag has not yielded convincing results, a separate tag was also developed. This relies on pre-formed gold nanoparticles (AuNPs) being introduced into the cell and binding to a specific protein. The obvious drawback is that not every protein will be labelled, and there may be some background mislabelling. However, the ligand used has very high affinity for the target protein, intracellular delivery has been demonstrated, and since the AuNP+ligand is so small penetration should be much greater than antibody or Fab′ labelling.

Battaglia et al. have designed polymersomes that are non-cytotoxic and can be used for rapid and efficient intracellular delivery of DNA [46] and fluorescent compounds [47]. Formation of polymersomes (composed of PMPC25-PDPA70 copolymer) and encapsulation of cargo occurs at pH 7.5. These polymersomes are taken up by cells *via* endocytosis (Fig. 5.18) before acidification of the endocytic compartment triggers the disassembly of the polymersome coat and release of cargo. The rapid increase in osmotic pressure results in rupture of the endocytic membrane and escape of cargo into the cytoplasm.

Clackson et al. have developed a mutant of FKBP12 (FKBP12(F36V)) that has picomolar affinity for a novel ligand, SLF′ [49]. SLF′ is membrane permeable and has been covalently attached to numerous fluorescent molecules [50, 51], whilst

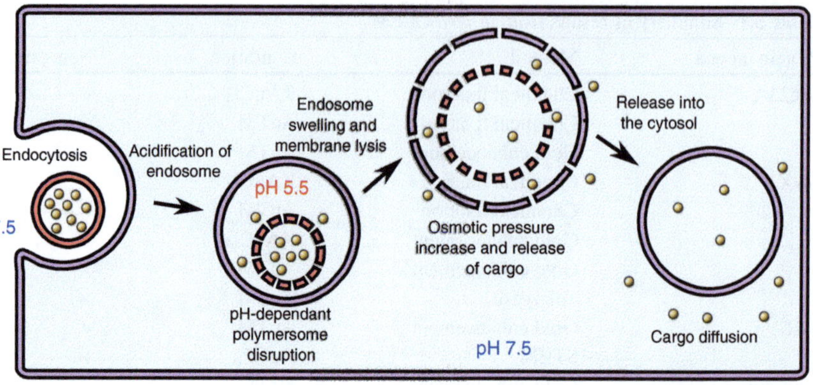

Fig. 5.18 Intracellular delivery of membrane-impermeable particles using pH-sensitive polymersomes. As the endosome is acidified by the cell the PMPC-PDPA block co-polymers become charged, disrupting the polymersome structure and releasing the contents into the endosome. As the osmotic pressure increases the endosome ruptures, releasing the contents into the cytoplasm (modified from [48])

FKBP12(F36V) has been genetically fused to a number of proteins within cells [50]. This has enabled fluorescent labelling of specific proteins within live cells.

Genetically fusing FKBP12(F36V) to GFP and a protein of interest (POI), combined with these two separate strategies for cellular delivery and *in vivo* protein labelling, will yield a fluorescent and electron-dense probe suitable for use in CLEM. SLF′ can be functionalised with monovalent 1.4 nm nanogold. Nanogold conjugated to antibodies has been used previously *in vivo* [52, 53] and showed low cytotoxicity. This will be encapsulated within the polymersome using a commercially available system and incubated with cells expressing POI-GFP-FKBP (Fig. 5.19). The 1.4 nm nanogold can be enhanced using post-fixation autometallography to 5 nm for imaging in the TEM using either conventional autometallography as described in this Thesis, or during freeze substitution after HPF [54]. Direct visualisation of the 1.4 nm nanogold using STEM may even be possible [55].

This project is in its early stages. So far, intracellular delivery of rhodamine dyes using the polymersome technology has been demonstrated (Fig. 5.20). Cells were stable after >24 h with the dye.

EGFP-FKBP was cloned next to a number of different proteins to test how well the insertions are tolerated. As with EGFP-MT2, both intracellular and extracellular proteins were chosen. Tetherin has been tagged to test whether the probe will bind without intracellular delivery required, as well as SNX1 and Sec23A. The insertions were all well tolerated, as judged by fluorescence microscopy. Timecourses of cells expressing each fusion protein revealed functional and correct transport (not shown). EGFP-FKBP was also subcloned adjacent to the bacterial transport protein YidC, which is found in the outer leaflet on the inner membrane [56]. After tagging, the bacteria should be thin enough to be frozen by rapid plunge freezing into liquid ethane. This will allow the probe to be tested as a tag for cryoET.

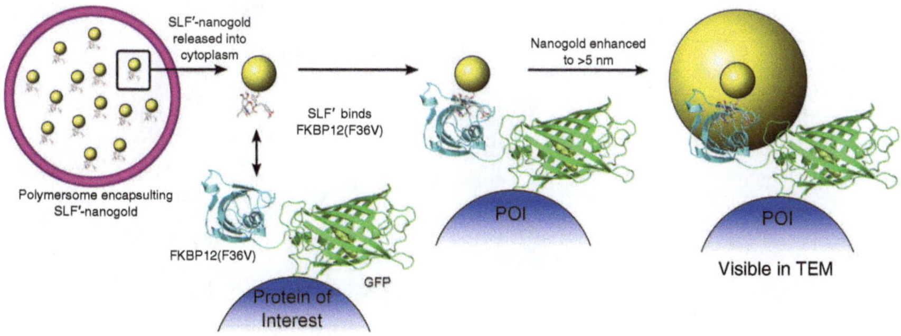

Fig. 5.19 Intracellular targeting of nanogold to GFP to yield a fluorescent and electron-dense probe suitable for CLEM. SLF′ covalently attached to 1.4 nm nanogold is encapsulated within a pH-sensitive polymersome and endocytosed by the cell. Following release into the cytosol this binds rapidly with pM affinity to the POI-EGFP-FKBP construct following release of the probe into the cytoplasm

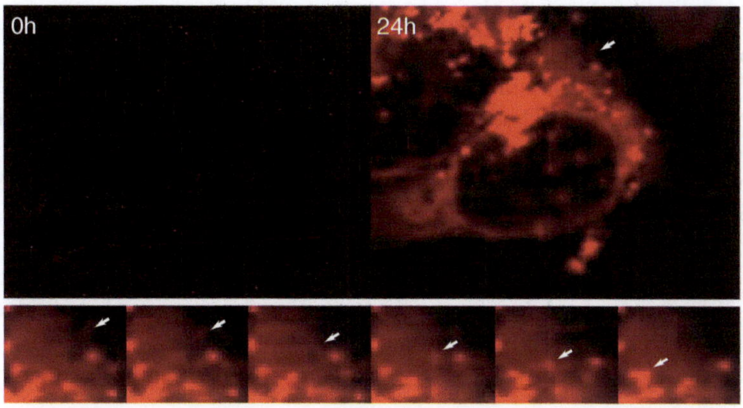

Fig. 5.20 Intracellular delivery of rhodamine dye. Polymersomes containing rhodamine were incubated with HeLa cells for 24 h (*upper row*). After 24 h, cells are clearly labelled both in vesicular bodies and the cytoplasm. A timecourse shows a magnified region of the cell (*arrow, upper row*) with a polymersome being endocytosed (*arrow, bottom row*) and delivered to an early endosome

Finally, the only impediment to this project is the chemical synthesis of SLF′. This is in the final stages of development but, unfortunately, was not ready for testing prior to completion of this Thesis. As soon as SLF′ is synthesised it will be conjugated to a fluorescein derivative and incubated with cells expressing mCherry-FKBP-SNX1 and mCherry-FKBP-Tetherin to assess binding of the ligand to proteins by colocalisation studies in the LM. Fluorescein-SLF′ is able to cross the plasma membrane directly since it is small and uncharged [50]. The mCherry fusion proteins were synthesised by simply digesting out EGFP and inserting mCherry, as the design of the vector intended. Fusions to mCherry were chosen so that fluorescein-SLF′ could be used,

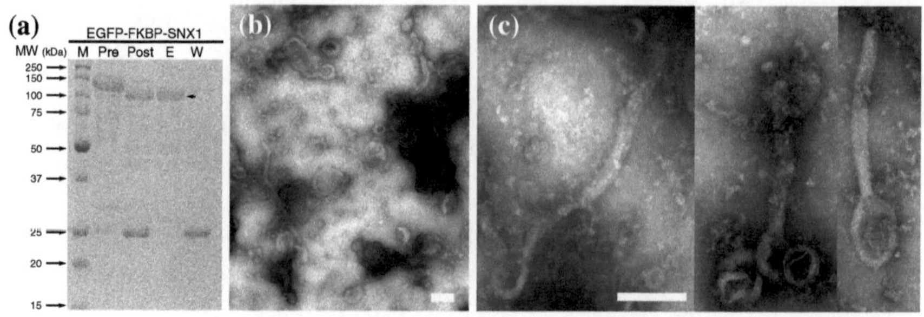

Fig. 5.21 EGFP-FKBP-SNX1 tubulates liposomes *in vitro*. **a** Purification of EGFP-FKBP-SNX1. *Arrow* indicates purified EGFP-FKBP-SNX1 in the eluate at 101.9 kDa. *M* Marker; *Pre* Precut EGFP-FKBP-SNX1; *Post* Postcut EGFP-FKBP-SNX1; *E* Eluate; *W* Wash. **b** Overview of liposomes with numerous tubules visible. **c** Example of tubules formed. Scale bars for (**a**) and (**b**) are both 200 nm

since the chemical synthesis of both nanogold-SLF′ and fluorescein-SLF′ share a common derivative of SLF′.

When the dual-colour colocalisation experiments are successful, nanogold will be conjugated to SLF′ and bound to purified EGFP-FKBP-SNX1 and the tubulation assay described above repeated. EGFP-FKBP-SNX1 was purified using the PreScission protease system, and revealed a band at the predicted weight of 101.9 kDa (Fig. 5.21a). It has already been demonstrated that purified EGFP-FKBP-SNX1 is able to tubulate liposomes *in vitro* (Fig. 5.21b, c). Indeed, the tubulation was much more efficient than when EMS1 was used, probably because FKBP does not contain ~30 % cysteine and therefore will not oxidise to form disulphide bridges, so cryoTEM of tubulated liposomes is a distinct possibility.

5.6 Discussion

Purification of His_6-tagged EGFP-MT2 has been optimised. MALDI MS has shown a minimal binding of ~29 Au^+ ions when incubated with an equimolar concentration of AuTM:Cysteine residues, indicating cooperative metal-binding by MT. A similar stoichiometry for $KAuCl_4$ was also found, albeit with a greatly reduced signal to noise ratio. DLS of purified protein in the presence of AuTM indicates a change in the folding of the protein, although it actually results in a decrease in the size. This may be due to the gold causing the protein to fold tighter around the gold-thiolate core. DLS of protein incubated with $KAuCl_4$ and $CdCl_2$ indicated that the metals were becoming reduced in some of the buffers used, resulting in Tris·HCl being used for all *in vitro* characterisation.

Liposomes encapsulating EGFP or EGFP-MT2 were produced by sonication and extrusion and imaged in the TEM. Liposomes ruptured in the vacuum of the microscope, although after incubation with $KAuCl_4$ electron dense nanoparticles were visualised proximally to the liposomes. Genetically fusing EGFP-MT2 to SNX1, a membrane tubulating protein, allowed the localisation of the MT2 domain to tubulated liposomes in the TEM. Diffuse electron dense patches were visible in both the presence and absence of stain. However, the MT2 domain appeared to impede the formation of tubules, perhaps due to the high concentration of cysteine residues increasing the likelihood of disulphide bonds forming and inhibiting protein function. Indeed, after synthesising increasing concatenations of MTn ($n = 4, 6, 8$, or 10), these were found to be non-purifiable, even under severe denaturing conditions. Similarly, Nishino et al. reported that GroEL tagged with MT4 or MT5 did not oligomerise correctly, and attributed this to steric hindrance due to the MT domain [3].

A bioinformatics approach was used to design and synthesise an MT2 domain with reduced repeating DNA sequences suitable for homologous recombination and subcloning within eukaryotes. This was subcloned into pEGFP-C1, a common commercial vector for EGFP expression in eukaryotic cell lines. The whole EGFP-MT2 cassette was subcloned adjacent to Sec23A, SNX1, α-tubulin, transferrin receptor, and Tetherin. The EGFP-MT2 domain did not perturb the function of any of the proteins with the exception of α-tubulin, as demonstrated by immunofluorescence and live-cell fluorescence imaging. Visualising the protein in both LM and TEM was achieved with the use of CelLocate coverslips, which enable the tracking back of specific cells. After serial sectioning, optical sections could be aligned based on visible features and cell shape. This allowed the identification of areas where EGFP-MT2 should be visible. However, only once was an electron dense particle seen. In an effort to increase the visibility of any nanoparticles that had formed, both immunoEM using the Tokuyasu technique and autometallography with gold compounds was attempted. Neither of these techniques yielded any electron dense particles that were visible solely at the location of interest. It appears that intracellular delivery of heavy-metal compounds to MT2 to form electron-dense particles *in vivo* is not compatible with live cell CLEM, although some promise has been shown here as an *in vitro* tag.

Protein tags based on the MT domain may have some uses in certain circumstances but, as this Thesis has gone some way to showing, are not suitable for general purpose tagging in the same way that GFP is for LM. Indeed, in the recent paper by Wang et al. [57] one of the authors, Dr. C. P. Mercogliano, also developed the first MT2 probes [1, 15] and filed a patent for its use as a "clonable" probe for EM, but states in the paper "…the heavy metals required for labelling are toxic for cells, and it remains unclear if they can be transported into cells with the required efficiency". Transport of heavy metals into cells could be increased using polymersome delivery (discussed in Sect. 5.5 and [48]), although the metal-catalysed oxidative stress may be too great for the cell. As metals are reduced, the resulting redox reaction leads to oxidation of cellular molecules or the production of oxygen free radicals [58, 59].

Another drawback to simple metal chelation by MT is not affinity for metals but non-specific binding. A more suitable method for CLEM may be to use a protein that forms electron dense nanoparticles enzymatically to avoid the need for toxic levels of metal ions. Unfortunately, no ideal proteins have been identified thus far, although there are a couple of systems which may be worth pursuing. Notably, ferritin has

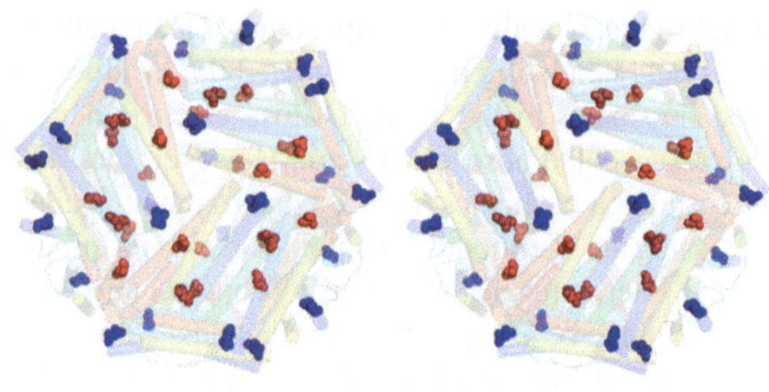

Fig. 5.22 Ferritin termini are distant in space. Stereoview showing N- and C- terminal residues, coloured *blue* and *red*, respectively. Backbone cartoons are coloured from *blue* through *red* accordingly. PDB code 1EUM

shown use as a tag for soluble intracellular proteins that is visible by TEM after incubation with iron compounds [57]. It may be that mutating ferritin to bind other, more electron dense, substrates could improve the tag [60, 61]. Alternatively, cadmium selenide (CdSe) can be mineralised within the ferritin cage [62]. As described in Sect. 1.5, CdSe forms the core of QDs, which may mean that fluorescent and electron dense particles may be grown *in situ*. However, the main drawback is that every ferritin subunit is attached to the protein of interest. This obligate oligomerisation means that in Wang et al. there are 24 subunits of each protein attached to each particle visible in the TEM [57]. It may be possible to mutate ferritin to form a monomer by concatenation of all 24 subunits, although this would yield a single protein domain of 466 kDa and fusion of the termini will be difficult due to their separation in space (Fig. 5.22).

A system that has not yet been explored is the magnetosomes of magnetotactic bacteria [63]. These are formed by an unidentified mechanism inside intracellular vesicles but most likely require enzymatic deposition and reduction to form. Isolation and manipulation of this protein may allow deposition of metal ions at specific loci.

Finally, in an attempt to avoid the pitfalls of using MT as an electron dense protein, i.e., high local concentrations of cysteine, a separate method was devised based on *in vivo* delivery of nanogold. Preliminary results are encouraging, the protein fusion does not interfere with protein function and the delivery system can reliably deliver fluorescent molecule into the cytoplasm. With the synthesis of the ligand, these systems can be amalgamated and a live-cell CLEM experiment attempted.

References

1. C.P. Mercogliano, D.J. DeRosier, Concatenated metallothionein as a clonable gold label for electron microscopy. J. Struct. Biol. **160**(1), 70–82 (2007)
2. J.H. Kagi, B.L. Valee, Metallothionein: a cadmium- and zinc-containing protein from equine renal cortex. The J. Biol. Chem. **235**, 3460–3465 (1960)
3. Y. Nishino, T. Yasunaga, A. Miyazawa, A genetically encoded metallothionein tag enabling efficient protein detection by electron microscopy. J. Electron Microsc. **56**(3), 93–101 (2007)
4. Y. Fukunaga, A. Hirase, H. Kim, N. Wada, Y. Nishino, A. Miyazawa, Electron microscopic analysis of a fusion protein of postsynaptic density-95 and metallothionein in cultured hippocampal neurons. J. Electron Microsc. **56**(4), 119–129 (2007)

5. W. Zhong, Reaction of a platinum(IV) complex with native Cd, Zn-metallothionein in vitro. J. Inorg. Biochem. **66**(3), 179–185 (1997)
6. J. Bongers, Platinum(II) binding to metallothioneins. J. Inorg. Biochem. **34**(1), 55–62 (1988)
7. H. Sun, H. Li, I. Harvey, P.J. Sadler, Interactions of bismuth complexes with metallothionein(II). J. Biol. Chem. **274**(41), 29094–29101 (1999)
8. J. Laib, F.C. Shaw, D.H. Petering, M.K. Eidsness, R.C. Elder, J.S. Garvey, Formation and characterization of aurothioneins: Au, Zn, Cd-thionein, Au, Cd-thionein, and (thiomalato-Au)x-thionein. Biochemistry **24**(8), 1977–1986 (1985)
9. G. Schmitz, D.T. Minkel, D. Gingrich, C.F. Shaw, The binding of Gold(I) to metallothionein. J. Inorg. Biochem. **12**(4), 293–306 (1980)
10. E. Diestra, B. Cayrol, V. Arluison, C. Risco, Cellular electron microscopy imaging reveals the localization of the Hfq protein close to the bacterial membrane. PLoS ONE **4**(12), e830+ (2009)
11. E. Diestra, J. Fontana, P. Guichard, S. Marco, C. Risco, Visualization of proteins in intact cells with a clonable tag for electron microscopy. J. Struct. Biol. **165**(3), 157–168 (2009)
12. C. Gammons, Y. Yu, A. Williamsjones, The disproportionation of gold(I) chloride complexes at 25 to 200°C. Geochim. Cosmochim. Acta **61**(10), 1971–1983 (1997)
13. X. Lu, H.Y. Tuan, B.A. Korgel, Y. Xia, Facile synthesis of gold nanoparticles with narrow size distribution by using AuCl or AuBr as the precursor. Chemistry **14**(5), 1584–1591 (2008)
14. A.T. Miles, G.M. Hawksworth, J.H. Beattie, V. Rodilla, Induction, regulation, degradation, and biological significance of mammalian metallothioneins. Crit. Rev. Biochem. Mol. Biol. **35**(1), 35–70 (2000)
15. C.P. Mercogliano, D.J. Derosier, Gold nanocluster formation using metallothionein: mass spectrometry and electron microscopy. J. Mol. Biol. **355**(2), 211–223 (2006)
16. J.H. Kaegi, A. Schaeffer, Biochemistry of metallothionein. Biochemistry **27**(23), 8509–8515 (1988)
17. J. Carlton, M. Bujny, B.J. Peter, V.M.J. Oorschot, A. Rutherford, H. Mellor, J. Klumperman, H.T. McMahon, P.J. Cullen, Sorting nexin-1 mediates tubular endosome-to-TGN transport through coincidence sensing of high- curvature membranes and 3-phosphoinositides. Curr. Biol. **14**(20), 1791–1800 (2004)
18. P.J. Cullen, Endosomal sorting and signalling: an emerging role for sorting nexins. Nat. Rev. Mol. Cell Biol. **9**(7), 574–582 (2008)
19. A. Frost, R. Perera, A. Roux, K. Spasov, O. Destaing, E.H. Egelman, P. De Camilli, V.M. Unger, Structural basis of membrane invagination by F-BAR domains. Cell **132**(5), 807–817 (2008)
20. J.R.T. van Weering, P. Verkade, P.J. Cullen, SNX-BAR proteins in phosphoinositide-mediated, tubular-based endosomal sorting. Semin. Cell Dev. Biol. (2009)
21. J.R.T. van Weering, P. Verkade, P.J. Cullen, SNX-BAR proteins in phosphoinositide-mediated, tubular-based endosomal sorting. Semin. Cell Dev. Biol. **21**(4), 371–380 (2010)
22. M. Welch, A. Villalobos, C. Gustafsson, J. Minshull, You're one in a googol: optimizing genes for protein expression. J. R. Soc. Interface **6**(Suppl 4), S467–S476 (2009)
23. H. Hughes, D. Stephens, Assembly, organization, and function of the COPII coat. Histochem. Cell Biol. **129**(2), 129–151 (2008)
24. J. Löwe, H. Li, K.H. Downing, E. Nogales, Refined structure of alpha beta-tubulin at 3.5 A resolution. J. Mol. Biol. **313**(5), 1045–1057 (2001)
25. A. Dautry-Varsat, A. Ciechanover, H.F. Lodish, pH and the recycling of transferrin during receptor-mediated endocytosis. PNAS **80**(8), 2258–2262 (1983)
26. S. Kupzig, V. Korolchuk, R. Rollason, A. Sugden, A. Wilde, G. Banting, Bst-2/HM1.24 is a raft-associated apical membrane protein with an unusual topology. Traffic (Copenhagen, Denmark) **4**(10), 694–709 (2003)
27. P.B. Schiff, S.B. Horwitz, Taxol stabilizes microtubules in mouse fibroblast cells. PNAS **77**(3), 1561–1565 (1980)
28. D.J. Stephens, N. Lin-Marq, A. Pagano, R. Pepperkok, J.P. Paccaud, COPI-coated ER-to-Golgi transport complexes segregate from COPII in close proximity to ER exit sites. J. Cell Sci. **113**(Pt 12), 2177–2185 (2000)

29. D.J. Stephens, De novo formation, fusion and fission of mammalian COPII-coated endoplasmic reticulum exit sites. EMBO reports **4**(2), 210–217 (2003)
30. S. Fath, J.D. Mancias, X. Bi, J. Goldberg, Structure and organization of coat proteins in the COPII cage. Cell **129**(7), 1325–1336 (2007)
31. J.S. Bonifacino, B.S. Glick, The mechanisms of vesicle budding and fusion. Cell **116**(2) 153–166 (2004)
32. N. Nakamura, C. Rabouille, R. Watson, T. Nilsson, N. Hui, P. Slusarewicz, T.E. Kreis, G. Warren, Characterization of a cis-Golgi matrix protein, GM130. J. Cell Biol. **131**(6 Pt 2), 1715–1726 (1995)
33. J. Shorter, G. Warren, A role for the vesicle tethering protein, p115, in the post-mitotic stacking of reassembling Golgi cisternae in a cell-free system. J. Cell Biol. **146**(1), 57–70 (1999)
34. A. Diao, L. Frost, Y. Morohashi, M. Lowe, Coordination of Golgin tethering and SNARE assembly: GM130 binds Syntaxin 5 in a p115-regulated manner. J. Biol. Chem. **283**(11), 6957–6967 (2008)
35. C. Alvarez, R. Garcia-Mata, H.P. Hauri, E. Sztul, The p115-interactive proteins GM130 and giantin participate in endoplasmic reticulum-Golgi traffic. J. Biol. Chem. **276**(4), 2693–2700 (2001)
36. B. Kleizen, Protein folding and quality control in the endoplasmic reticulum. Curr. Opin. Cell Biol. **16**(4), 343–349 (2004)
37. W. He, C. Kivork, S. MacHinani, M. Morphew, A. Gail, D. Tesar, N. Tiangco, J. McIntosh, P. Bjorkman, A freeze substitution fixation-based gold enlarging technique for EM studies of endocytosed Nanogold-labeled molecules. J. Struct. Biol. **160**(1), 103–113 (2007)
38. K.T. Tokuyasu, A technique for ultracryotomy of cell suspensions and tissues. J. Cell Biol. **57**(2), 551–565 (1973)
39. J.R. van Weering, E. Brown, T.H. Sharp, J. Mantell, P.J. Cullen, P. Verkade, Intracellular membrane traffic at high resolution. Methods Cell Biol. **96**, 619–648 (2010)
40. W.R. Bernhard, M. Vasak, J.H.R. Kagi, Cadmium binding and metal cluster formation in metallothionein: a differential modification study. Biochemistry **25**(8), 1975–1980 (1986)
41. E. Brown, P. Verkade, The use of markers for correlative light electron microscopy. Protoplasma **244**(1–4), 91–97 (2010)
42. E. Brown, J. Mantell, D. Carter, G. Tilly, P. Verkade, Studying intracellular transport using high-pressure freezing and Correlative Light Electron Microscopy. Semin. Cell Dev. Biol. **20**(8), 910–919 (2009)
43. D. Perez-Caballero, T. Zang, A. Ebrahimi, M.W. McNatt, D.A. Gregory, M.C. Johnson, P.D. Bieniasz, Tetherin inhibits HIV-1 release by directly tethering virions to cells. Cell **139**(3), 499–511 (2009)
44. S.J.D. Neil, T. Zang, P.D. Bieniasz, Tetherin inhibits retrovirus release and is antagonized by HIV-1 Vpu. Nature **451**(7177), 425–430 (2008)
45. J. Hammonds, J.-J. Wang, H. Yi, P. Spearman, Immunoelectron microscopic evidence for Tetherin/BST2 as the physical bridge between HIV-1 virions and the plasma membrane. PLoS Pathog. **6**(2), e1000749+ (2010)
46. H. Lomas, I. Canton, S. Macneil, J. Du, S.P. Armes, A.J. Ryan, A.L. Lewis, G. Battaglia, Biomimetic pH sensitive polymersomes for efficient DNA encapsulation and delivery. Adv. Mater. **19**(23), 4238–4243 (2007)
47. H. Lomas, M. Massignani, K.A. Abdullah, I. Canton, C.L. Presti, S. Macneil, J. Du, A. Blanazs, J. Madsen, S.P. Armes, A.L. Lewis, G. Battaglia, Non-cytotoxic polymer vesicles for rapid and efficient intracellular delivery. Faraday Discuss. **139**, 143–159 (2008)
48. C. Lopresti, H. Lomas, M. Massignani, T. Smart, G. Battaglia, Polymersomes: nature inspired nanometer sized compartments. J. Mater. Chem. **19**(22), 3576–3590 (2009)
49. T. Clackson, W. Yang, L.W. Rozamus, M. Hatada, J.F. Amara, C.T. Rollins, L.F. Stevenson, S.R. Magari, S.A. Wood, N.L. Courage, X. Lu, F. Cerasoli, M. Gilman, D.A. Holt, Redesigning an FKBP-ligand interface to generate chemical dimerizers with novel specificity. PNAS **95**(18), 10437–10442 (1998)

50. K.M. Marks, P.D. Braun, G.P. Nolan, A general approach for chemical labeling and rapid, spatially controlled protein inactivation. PNAS **101**(27), 9982–9987 (2004)
51. M. Robers, P. Pinson, L. Leong, R.H. Batchelor, K. R. Gee, T. Machleidt, Fluorescent labeling of proteins in living cells using the FKBP12 (F36V) tag. J. Int. Soc. Anal. Cytol. A **75**(3), 207–224 (2009)
52. I. Kireev, M. Lakonishok, W. Liu, V.N. Joshi, R. Powell, A.S. Belmont, In vivo immunogold labeling confirms large-scale chromatin folding motifs. Nat. Methods **5**(4), 311–313 (2008)
53. Y. Hu, I. Kireev, M. Plutz, N. Ashourian, A.S. Belmont, Large-scale chromatin structure of inducible genes: transcription on a condensed, linear template. J. Cell Biol. **185**(1), 87–100 (2009)
54. M. Morphew, W. He, P.J. Bjorkman, J.R. McIntosh, Silver enhancement of Nanogold particles during freeze substitution for electron microscopy. J. Microsc. **230**(Pt 2), 263–267 (2008)
55. A. Sousa, M.A. Aronova, Y. Kim, L. Dorward, G. Zhang, R.D. Leapman, Limits of detection of ultrasmall gold labels in biological specimens by STEM tomography. Microscopy and Microanalysis **13**(Suppl. S02), 1328–1329 (2007)
56. J.C. Samuelson, M. Chen, F. Jiang, I. Moller, M. Wiedmann, A. Kuhn, G.J. Phillips, R.E. Dalbey, YidC mediates membrane protein insertion in bacteria. Nature **406**(6796), 637–641 (2000)
57. Q. Wang, C.P. Mercogliano, J. Löwe, A ferritin-based label for cellular electron cryotomography. Structure **19**(2), 147–154 (2011)
58. S. Stohs, Oxidative mechanisms in the toxicity of metal ions. Free Radical Biol. Med. **18**(2), 321–336 (1995)
59. F. Reith, B. Etschmann, C. Grosse, H. Moors, M.A. Benotmane, P. Monsieurs, G. Grass, C. Doonan, S. Vogt, B. Lai, G. Martinez-Criado, G.N. George, D.H. Nies, M. Mergeay, A. Pring, G. Southam, J.A. Brugger, Mechanisms of gold biomineralization in the bacterium Cupriavidus metallidurans. PNAS (2009)
60. R. Fan, S.W. Chew, V.V. Cheong, B.P. Orner, Fabrication of gold nanoparticles inside unmodified horse spleen apoferritin. Small **6**(14), 1483–1487 (2010)
61. Z. Wang, Y. Takezawa, H. Aoyagi, S. Abe, T. Hikage, Y. Watanabe, S. Kitagawa, T. Ueno, Definite coordination arrangement of organometallic palladium complexes accumulated on the designed interior surface of apo-ferritin. Chem. Commun. **47**(1), 170–172 (2011)
62. K. Iwahori, T. Morioka, I. Yamashita, The optimization of CdSe nanoparticles synthesis in the apoferritin cavity. Phys. Status Solidi A **203**(11), 2658–2661 (2006)
63. A. Scheffel, M. Gruska, D. Faivre, A. Linaroudis, J.M. Plitzko, D. Schuler, An acidic protein aligns magnetosomes along a filamentous structure in magnetotactic bacteria. Nature **440**(7080), 110–114 (2005)

Concluding Remarks

The generation of the first high-resolution structure of a de novo designed self-assembling peptide fibre has been achieved using a combination of cryoTEM, 2D-crystal image analysis and helical reconstruction. With this information and an X-ray crystal structure of a related peptide complex, an all-atom model for the SAFs has been generated. This structure has been shown to be located at a minimum on the energy landscape of the component peptides by molecular dynamics simulations. In terms of supramolecular peptide assembly, these new structural insights, and the recently determined assembly pathway [1], mean that the SAFs represent the most well understood designed, peptide-based, fibre-forming system presented to date. In addition, a key finding of this Thesis is that coiled-coil fibrils can be remodelled, in this case compressed, to allow better packing within a larger supramolecular structure. This may shed light on the assembly and packing of similar fibres, including the intermediate filaments, which have eluded structural definition to high resolution. Finally, the structures and models presented offer a scaffold for engineering the SAFs towards to realising their application in bionanotechnology through rational materials design.

Secondly, the design of genetically encoded tags for CLEM has been described. Results published separately by others during the work described in this Thesis demonstrated the potential viability of MT, MT2 and MT3 tags for localisation both *in vivo* and *in vitro* [2–6]. However, only one of these studies described its use in eukaryotic cells [4], and the use of toxic cadmium precludes its widespread use. In an effort to obtain a tag for general use for CLEM, MT2 was genetically fused to EGFP and tested in a number of *in vitro* and *in vivo* assays. This demonstrated the visibility of the tag *in vitro* when encapsulated in liposomes and potentially along protein mediated membrane tubules. However, *in vivo* the tag has not been reliably visualised. Indeed, despite the use of chemical fixation, gold enhancement, Tokuyasu immunoEM and STEM, the tag has not demonstrated the required localisation properties. Since the publication of the viability of the MT2 tag in 2007 [2] has not generated widespread use of the tag, together with my own results, I conclude that

T. H. Sharp, *Biomolecular Imaging at High Spatial and Temporal Resolution In Vitro and In Vivo*, Springer Theses, DOI: 10.1007/978-3-319-02159-1,

the delivery of sufficient numbers of metal ions to the MT tag precludes the use of MT as an *in vivo* localisation marker for EM.

Separately, the use of a newly developed system for the intracellular delivery of particles has been utilised for the potential delivery of monofunctionalised nanogold. Together with an orthogonal protein-ligand pair, the specific intracellular delivery of an electron dense marker for TEM and cryoTEM has show promising initial results. Synthesis of the ligand should yield an adaptable intracellular marker for both cryoTEM and TEM.

References

1. E.H. Bromley, K.J. Channon, P.J. King, Z.N. Mahmoud, E.F. Banwell, M.F. Butler, M.P. Crump, T.R. Dafforn, M.R. Hicks, J.D. Hirst, A. Rodger, D.N. Woolfson, Assembly pathway of a designed alpha-helical protein fiber. Biophys. J. **98**(8), 1668–1676 (2010)
2. C.P. Mercogliano, D.J. DeRosier, Concatenated metallothionein as a clonable gold label for electron microscopy. J. Struct. Biol. **160**(1), 70–82 (2007)
3. Y. Nishino, T. Yasunaga, A. Miyazawa, A genetically encoded metallothionein tag enabling efficient protein detection by electron microscopy. J. Electron Microsc. **56**(3), 93–101 (2007)
4. Y. Fukunaga, A. Hirase, H. Kim, N. Wada, Y. Nishino, A. Miyazawa, Electron microscopic analysis of a fusion protein of postsynaptic density-95 and metallothionein in cultured hippocampal neurons. J. Electron Microsc. **56**(4), 119–129 (2007)
5. E. Diestra, J. Fontana, P. Guichard, S. Marco, C. Risco, Visualization of proteins in intact cells with a clonable tag for electron microscopy. J. Struct. Biol. **165**(3), 157–168 (2009)
6. E. Diestra, B. Cayrol, V. Arluison, C. Risco, Cellular electron microscopy imaging reveals the localization of the Hfq protein close to the bacterial membrane. PLoS ONE textbf4(12), e8301+ (2009).

Appendix A
Amino Acid Structures

See Fig. A.1.

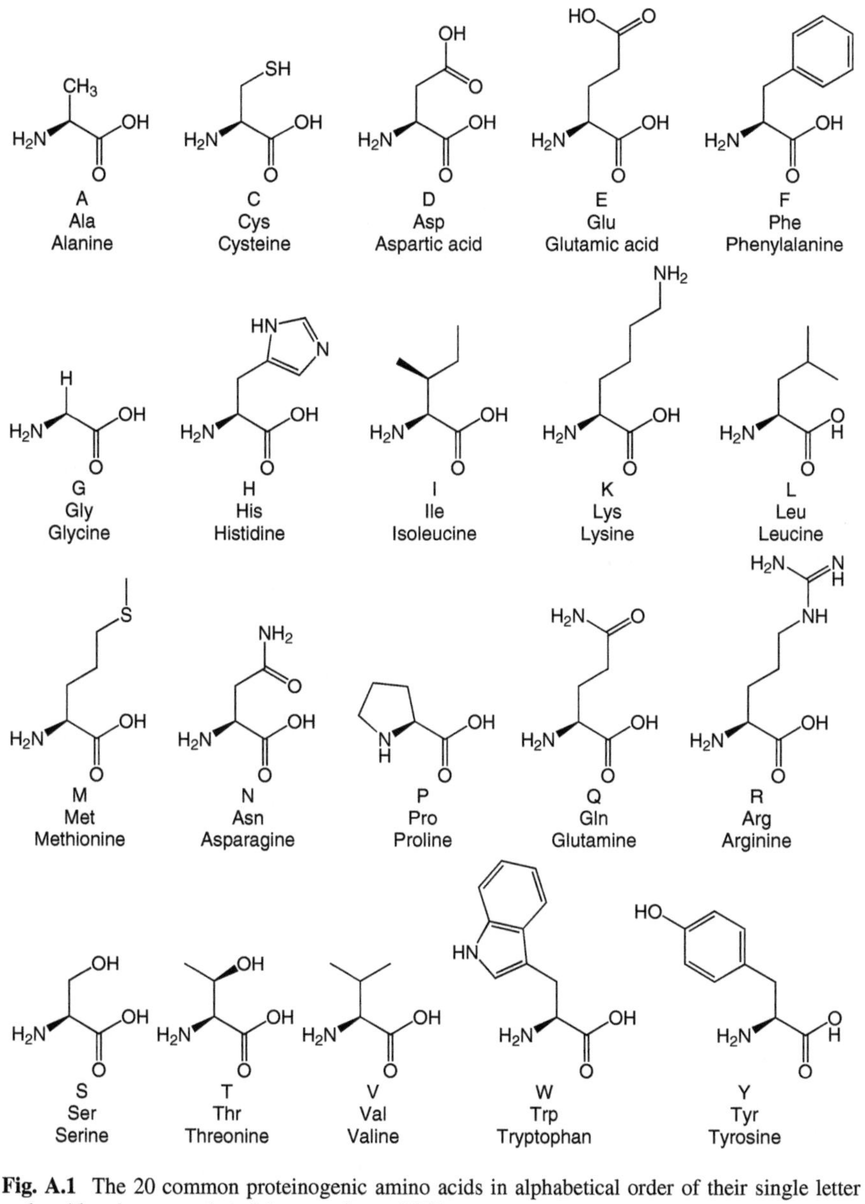

Fig. A.1 The 20 common proteinogenic amino acids in alphabetical order of their single letter codes. Also shown are their 3-letter codes and full name

T. H. Sharp, *Biomolecular Imaging at High Spatial and Temporal Resolution In Vitro and In Vivo*, Springer Theses, DOI: 10.1007/978-3-319-02159-1,

Appendix B
Vectors, Primers and Recipes

B.1 Vectors

Described below are the vectors synthesised for this Thesis.

B.1 Prokaryotic Expression Vectors

pET-3d-MT2
pET-45-b(+)-EGFP
pET-45-b(+)-EGFP-MT2
pET-45-b(+)-EGFP-MT4
pET-45-b(+)-EGFP-MT6
pET-45-b(+)-EGFP-MT8
pET-45-b(+)-EGFP-MT10
pET-45-b(+)-EGFP-SNX1
pET-45-b(+)-EGFP-MT2-SNX1
pET-45-b(+)-EGFP-FKBP-SNX1.

B.2 Eukaryotic Expression Vectors

pEGFP-MT2-Sec23A
pEGFP-MT2-SNX1
pEGFP-BsiWI-MT2-SNX1
pEGFP-MT2-Tubulin
pEGFP-MT2-TfR
pCR3.1-EGFP-MT2-Tetherin
pEGFP-FKBP-Sec23A

T. H. Sharp, *Biomolecular Imaging at High Spatial and Temporal Resolution In Vitro and In Vivo*, Springer Theses, DOI: 10.1007/978-3-319-02159-1,

pEGFP-FKBP-SNX1
pEGFP-FKBP-TfR
pCR3.1-EGFP-FKBP-Tetherin
pmCherry-FKBP-SNX1.

B.2 Primers

See Table B.1

B.3 Recipes

TBE buffer: 90 mM Tris Borate, 2 mM EDTA, pH 8.3.
Destaining solution: 10 % Glacial acetic acid, 50 % methanol in ddH_2O.
PBS: 3.8 mM NaH_2PO_4, 16.2 mM Na_2HPO_4, 150 mM NaCl, pH 7.4.
PBS-BSA: 1 % (v/v) acetylated BSA in PBS, pH 7.4.
PBS-Gly: 50 mM glycine in PBS, pH 7.4.
Solution E: 5 mM NaH_2PO_4, 100 mM NaCl, pH 5.5.

Table B.1 Described above are the primers in the order that they appear in the materials and methods chapter. Restriction endonuclease sites are underlined as described in the text

Primer name	Sequence (5′–3′)
His EGFP FWD	GATCCCAATTGGATGGTGAGCAAGGGCGAGG
His EGFP REV	TGAGCTCGAGATCTGAGTCC
T7 Prom	TAATACGACTCACTATAGGG
pET-3d-MT2 REV	AGAGGTCATGACACAGCACGTGCACTTGTCC
pET45-link-MT2	AATTTTGTTTAACTTTAATGTACAGATCTACCATGG
T7 Term	GCTAGTTATTGCTCAGCGG
TethGFP For	GCGCTACCTGCAGCAGGAATGGTGAGCAAGGGCGAGG
TethMT2 Rev	TCTGAGCCGCGGCTTGTACAGTGCGCAACAGG
FKBP FWD	CTTCGAAGACCAGTACGCGTACGGCACTAGTATGGCTTCTAGAGGAGTGC
FKBP REV	AGTCTGGTACGTCGTGTACAGCGCCGGTGATTCCAGTTTTAGAAGCTCCA

MIX
Papier aus verantwortungsvollen Quellen
Paper from responsible sources
FSC® C105338

If you have any concerns about our products,
you can contact us on
ProductSafety@springernature.com

In case Publisher is established outside the EU,
the EU authorized representative is:
Springer Nature Customer Service Center GmbH
Europaplatz 3, 69115 Heidelberg, Germany

Printed by Libri Plureos GmbH
in Hamburg, Germany